FEYNMAN
LECTURES ON
GRAVITATION

RICHARD P. FEYNMAN
FERNANDO B. MORINIGO · WILLIAM G. WAGNER

Edited by Brian Hatfield

With a Foreword by
John Preskill and Kip S. Thorne

Advanced Book Program

CRC Press
Taylor & Francis Group
Boca Raton London New York

CRC Press is an imprint of the
Taylor & Francis Group, an **informa** business

The Advanced Book Program

First published 2003 by Westview Press

Published 2018 by CRC Press
Taylor & Francis Group
6000 Broken Sound Parkway NW, Suite 300
Boca Raton, FL 33487-2742

CRC Press is an imprint of the Taylor & Francis Group, an informa business

Visit the Taylor & Francis Web site at
http://www.taylorandfrancis.com

and the CRC Press Web site at
http://www.crcpress.com

A Cataloging-in-Publication data record for this book is available
from the Library of Congress.

Cover design by Lynne Reed
Text design by Brian Hatfield
Set in 11-point Computer Modern Roman by Brian Hatfield

ISBN 13: 978-0-8133-4038-8 (pbk)

Contents

Foreword

John Preskill and Kip S. Thorne

During the 1962–63 academic year, Richard Feynman taught a course at Caltech on gravitation. Taking an untraditional approach to the subject, Feynman aimed the course at advanced graduate students and postdoctoral fellows who were familiar with the methods of relativistic quantum field theory—in particular, with Feynman-diagram perturbation theory in quantum electrodynamics. Two of the postdocs, Fernando B. Morinigo and William G. Wagner, wrote up notes for the course. These were typed, and copies were sold in the Caltech bookstore for many years.

The notes were not published, but they were widely distributed, and because of their unique insights into the foundations of physics, they had a significant influence on many of the people who read them. Morinigo and Wagner performed a great service by preserving so well this piece of Feynman's legacy. Now, thanks to the efforts of Brian Hatfield, the lectures are finally being published in this volume, and so are becoming readily available to a broader audience and to posterity. In preparing the notes for publication, Hatfield has corrected minor errors and improved the notation, but otherwise has adhered to the original typescript prepared by Morinigo and Wagner. (Only two brief passages were completely deleted.[1])

Feynman delivered 27 lectures in all, one each week over the full 1962–63 academic year. The class met in a tiny room with just two rows

---

[1] The deleted passages are an incorrect expression for the matter field action at the conclusion of §8.7 (the correct expression appears in §10.2), and some incorrect statements about the Newtonian theory of stellar stability in the third paragraph of §14.3.

of seats, on the third floor of Caltech's East Bridge Laboratory; no more than 15 people attended a typical lecture. (At least two of the students attending, James Bardeen and James Hartle, later made significant contributions of their own to gravitation theory.) The lectures were taped, but because they were highly informal, Morinigo and Wagner found it necessary to revise the material substantially in order to produce a readable set of notes. For the most part, Wagner worked on the mathematical presentation and Morinigo worked on the prose. The resulting notes were then reviewed by Feynman, he made various corrections and additions, and the notes were distributed to the students. The notes are pervaded by Feynman's spirit and are sprinkled with his jokes, but inevitably his idiosyncratic use of language has been only partially preserved.

Only 16 lectures are included in this book; these correspond, roughly, to the first 16 of the 27 that Feynman gave. Morinigo and Wagner prepared notes for all 27 lectures, but by the end of the academic year Feynman had reviewed and corrected only the first 11. Evidently he then became distracted by other projects and never returned to the task of editing the notes. Thus, only the notes for the first 11 lectures were distributed to the students during 1962–63 and were reproduced for sale by the Caltech bookstore in succeeding years.

In July 1971, a new reproduction of the notes was prepared for bookstore distribution, and Feynman authorized the inclusion of the next 5 lectures. The new lectures were prefaced by a disclaimer:

> The wide interest in these Lectures on Gravitation has led to a third reproduction. While this edition was being prepared, Professor Feynman kindly authorized the inclusion of the following five lectures. These lectures were intended to accompany the previous eleven in the 1962–63 notes, but were never satisfactorially (sic) edited and corrected for Professor Feynman to feel that they should be included.

> These lectures retain their rough form: except for minor errors corrected in transcription, they remain the same as they were eight years ago: Professor Feynman has not checked them. It is hoped that the reader will bear this in mind and see the following lectures as an account of what Professor Feynman was thinking at that time, rather than as an authoritative exposition of his work.

It seems true, indeed, that Feynman did not check the new lectures in detail. For example, Lecture 14 contains incorrect statements (discussed below) that he would have easily recognized as incorrect in 1971 (or even within a few weeks after giving the lecture) had he checked them. Thus, we urge readers to keep the above disclaimer in mind when reading Lectures 12–16.

Because Feynman never authorized the distribution of the Morinigo-Wagner notes for the last 11 lectures, they are omitted from this volume. These later lectures were mostly concerned with radiative corrections in quantum gravity and in Yang-Mills theory. We assume that Feynman did not want them distributed because he was dissatisfied with their content.

It is remarkable that concurrently with this course on gravitation, Feynman was also creating and teaching an innovative course in sophomore (second-year undergraduate) physics, a course that would become immortalized as the second and third volumes of *The Feynman Lectures on Physics* [Feyn 63a]. Each Monday Feynman would give his sophomore lecture in the morning and the lecture on gravitation after lunch. Later in the week would follow a second sophomore lecture and a lecture for scientists at Hughes Research Laboratories in Malibu. Besides this teaching load and his own research, Feynman was also serving on a panel to review textbooks for the California State Board of Education, itself a consuming task, as is vividly recounted in *Surely You're Joking, Mr. Feynman* [Feyn 85]. Steven Frautschi, who attended the lectures on gravitation as a young Caltech assistant professor, remembers Feynman later saying that he was "utterly exhausted" by the end of the 1962–63 academic year.

Feynman's course was never intended to be a comprehensive introduction to general relativity, and some of the lectures have become seriously dated. Much of the material in Lectures 7–12 is covered more systematically and in greater detail in other books. Why then, should the lectures be published now? There are at least three good reasons. First, nowhere else is there a comparable pedagogical account of an unusual approach to the foundations of general relativity that was pioneered by Feynman (among others). That approach, presented in lectures 3–6, develops the theory of a massless spin-2 field (the graviton) coupled to the energy-momentum tensor of matter, and demonstrates that the effort to make the theory self-consistent leads inevitably to Einstein's general relativity. (It is for this demonstration that the notes have become best known in the physics community.) Second, the notes contain a number of fascinating digressions and asides on the foundations of physics and other issues that make them enlightening and interesting to read. Third, the notes have historical value. At the time he taught the course, Feynman had been thinking hard for several years about fundamental problems in gravitation, and it is useful to have a record of his insights and viewpoint at that particular time. Some of his opinions seem prescient to us 32 years later, while others, naturally, seem naive or wrong. In some cases, his views were evolving rapidly as he taught the course. This is particularly true of the material in Lecture 14 on relativistic stars, about which we will say more below.

While these lectures are of special value for what they teach us of Feynman's viewpoints on gravitation, they are *not* a good place for a beginning student to learn the modern geometrical formulation of general relativity, or the computational tools and applications of the theory. Books such as Wald [Wald 84], Schutz [Schu 85], and Misner, Thorne, and Wheeler [MTW 73] do a much better job of this pedagogical task. Even the dogmatically nongeometrical viewpoint preferred by Feynman is presented more systematically and comprehensively by Weinberg [Wein 72]. But no other source contains Feynman's unique insights and approach to the foundations of the subject.

The lecture notes can be read at several different levels by people of varying backgrounds:

- To understand the lectures fully, readers should have advanced training in theoretical physics. Feynman assumes that his audience is familiar with the methods of quantum field theory, to the extent of knowing how to extract Feynman rules from an action and how to use the rules to calculate tree diagrams. However, these field theory methods are used heavily only in Lectures 2–4 and 16, and even in these chapters the key ideas can be grasped without such training. Moreover, the other lectures can be read more or less independently of these.

- Readers with a solid undergraduate training in physics should find the lectures largely intelligible, thanks to Feynman's pedagogical skill. However, such readers will have to be content with a heuristic grasp of some of the more technical portions.

- For admirers of Feynman who do not have a strong physics background, these lectures may also contain much of value—though to dig it out will require considerable skimming of the technical material that is interspersed with more down to earth insights and passages.

The remainder of this foreword, and a following section by Brian Hatfield, present a synopsis of the lectures and a discussion of how they relate to prior research on gravitation and to subsequent developments. Like the lectures themselves, our synopsis can be read at different levels. To assist readers who do not have advanced training in theoretical physics, we have marked with marginal grey bars some especially technical sections of the foreword that they may wish to skip or only browse.

Derivation of the Einstein field equation:

At the time of these lectures, Feynman was struggling to quantize gravity —that is, to forge a synthesis of general relativity and the fundamental principles of quantum mechanics. Feynman's whole approach to general relativity is shaped by his desire to arrive at a quantum theory of gravitation as straightforwardly as possible. For this purpose, geometrical subtleties seem a distraction; in particular, the conventional geometrical approach to gravitation obscures the telling analogy between gravitation and electrodynamics.

With hindsight, we can arrive at Maxwell's classical electrodynamics by starting with the observation that the photon is a massless spin-1 particle. The form of the quantum theory of a massless spin-1 particle coupled to charged matter is highly constrained by fundamental principles such as Lorentz invariance and conservation of probability. The self-consistent version of the quantum theory—quantum electrodynamics—is governed, in the classical limit, by Maxwell's classical field equations.

Emboldened by this analogy, Feynman views the quantum theory of gravitation as "just another quantum field theory" like quantum electrodynamics. Thus he asks in Lectures 1–6: can we find a sensible quantum field theory describing massless spin-2 quanta (gravitons) coupled to matter, in ordinary flat Minkowski spacetime? The classical limit of such a quantum theory should be governed by Einstein's general relativistic field equation for the classical gravitational field. Therefore, to ascertain the form of the classical theory, Feynman appeals to the features of the quantum theory that must underlie it. Geometrical ideas enter Feynman's discussion only through the "back door," and are developed primarily as technical tools to assist with the task of constructing an acceptable theory. So, for example, the Riemannian curvature tensor (a centerpiece of the conventional formulation of general relativity) is introduced by Feynman initially (§6.4) only as a device for constructing terms in the gravitational action with desired invariance properties. Not until §9.3 does Feynman reveal that the curvature has an interpretation in terms of the parallel transport of a tangent vector on a curved spacetime manifold.

One crucial feature of the quantum theory is that the massless spin-2 graviton has just two helicity states. Thus, the classical gravitational field must also have only two dynamical degrees of freedom. However, the classical field that corresponds to a spin-2 particle is a symmetric tensor $h_{\mu\nu}$ with ten components. Actually, four of these components h_{00}, h_{0i} (with $i = 1, 2, 3$) are nondynamical constrained variables, so we are left with six dynamical components h_{ij} to describe the two physical helicity states. It is because of this mismatch between the number of particle states and the number of field components that the quantum theory of gravitation, and hence also the corresponding classical theory, are highly constrained.

To resolve this counting mismatch, it is necessary for the theory to incorporate a redundancy, so that many different classical field configurations describe the same physical state. In other words, it must be a gauge theory. For the massless spin-2 field, the requisite gauge principle can be shown to be general covariance, which leads to Einstein's theory.

In Lecture 3, Feynman constructs the quadratic action of a massless spin-2 field that is linearly coupled to a conserved energy-momentum tensor. He comments on the gauge invariance of the resulting linear field equation in §3.7, and he remarks in §4.5 that one can infer the nonlinear self-couplings of the field by demanding the gauge invariance of the scattering amplitudes. But he does not carry this program to completion. (He notes that it would be hard to do so.) Instead, he uses a rather different method to arrive at Einstein's nonlinear classical field equation—a method that focuses on *consistency*. Because the linear field equation of the *free* massless spin-2 field necessarily has a gauge invariance (to remove the unwanted helicity states), generic modifications of this field equation (such as the modifications that arise when the spin-2 field is coupled to matter) do not admit any solutions. The new terms in the modified equation must respect a nontrivial *consistency condition*, which is essentially the requirement that the new terms respect the gauge symmetry. This consistency condition is sufficient, when pursued, to point the way toward Einstein's specific set of nonlinear couplings, and his corresponding nonlinear field equation.

In greater detail: the problem, as formulated in §6.2, is to find an action functional $F[h]$ for the spin-2 field $h_{\mu\nu}$ such that the gravitational field equation

$$\frac{\delta F}{\delta h_{\mu\nu}} = T^{\mu\nu} \tag{F.1}$$

is consistent with the equation of motion satisfied by the matter. Here $T^{\mu\nu}$ is the matter's energy-momentum tensor. Feynman finds a quadratic expression for F in Lecture 3, which yields a consistent linear field equation so long as the matter's energy-momentum is conserved (in the special relativistic sense), $T^{\mu\nu}{}_{,\nu} = 0$. The trouble is that, once the field $h_{\mu\nu}$ is coupled to the matter (so that matter acts as a source for $h_{\mu\nu}$), the equation of motion of the matter is modified by gravitational forces, and $T^{\mu\nu}{}_{,\nu}$ no longer vanishes. Thus, the field equation of $h_{\mu\nu}$ and the equation of motion of the matter are not compatible; the equations admit no simultaneous solutions. This is the consistency problem (of the linear theory).

By demanding that the field equation satisfied by $h_{\mu\nu}$ be compatible with the matter's equation of motion, Feynman infers that higher-order nonlinear corrections must be added to the action, F. The consistency requirement can be cast in the form of an invariance principle satisfied by the action, which is just invariance under general coordinate transformations. After that, Feynman's analysis is fairly conventional, and leads to the conclusion that the most general consistent field equation involving no more than two derivatives is the Einstein equation (with a cosmological constant).

The resulting nonlinear corrections have a pleasing physical interpretation. Without these corrections, gravity does not couple to itself. When nonlinear corrections of the required form are included, the source for the gravitational field (as viewed in flat Minkowski spacetime) is the *total* energy momentum, including the contribution due to the gravitational field itself. In other words, the (strong) principle of equivalence is satisfied. The conservation law obeyed by the energy momentum of the matter becomes Einstein's covariant one, $T^{\mu\nu}{}_{;\nu} = 0$, which in effect allows energy and momentum to be exchanged between matter and gravity.

We know from Feynman's comments at the 1957 Chapel Hill conference [DeWi 57] that by then he had already worked out the calculations described in Lectures 2–6. Murray Gell-Mann reports [Gell 89] that Feynman and he discussed quantum gravity issues during the Christmas break in 1954–55, and that Feynman had already made "considerable progress" by that time.

The claim that the only sensible theory of an interacting massless spin-2 field is essentially general relativity (or is well approximated by general relativity in the limit of low energy) is still often invoked today. (For example, one argues that since superstring theory contains an interacting massless spin-2 particle, it must be a theory of gravity.) In fact, Feynman was not the very first to make such a claim.

The field equation for a free massless spin-2 field was written down by Fierz and Pauli in 1939 [FiPa 39]. Thereafter, the idea of treating Einstein gravity as a theory of a spin-2 field in flat space surfaced occasionally in the literature. As far as we know, however, the first published attempt to *derive* the nonlinear couplings in Einstein's theory in this framework appeared in a 1954 paper by Suraj N. Gupta [Gupt 54]. Gupta noted that the action of the theory must obey a nontrivial consistency condition that is satisfied by general relativity. He did not, however, provide any detailed argument supporting the *uniqueness* of the Einstein field equation.

Roughly speaking, Gupta's argument proceeds as follows: We wish to construct a theory in which the "source" coupled to the massless spin-2 field $h_{\mu\nu}$ is the energy-momentum tensor, *including* the energy-momentum of the spin-2 field itself. If the source is chosen to be the energy-momentum tensor $^2T^{\mu\nu}$ of the free field theory (which is quadratic in h), then coupling this source to $h_{\mu\nu}$ induces a cubic term in the Lagrangian. From this cubic term in the Lagrangian, a corresponding cubic term $^3T^{\mu\nu}$ in the energy-momentum tensor can be inferred, which is then included in the source. This generates a quartic term $^4T^{\mu\nu}$, and so on. This iterative procedure generates an infinite series that can be summed to yield the full nonlinear Einstein equations. Gupta sketched this procedure, but did not actually carry it to completion. The first complete version of the argument (and an especially elegant one) was published by Deser in 1970 [Dese 70].[2] Deser also noted that Yang-Mills theory can be derived by a similar argument.

Some years before Gupta's work, Robert Kraichnan, then an 18-year-old undergraduate at M.I.T., had also studied the problem of deriving general relativity as a consistent theory of a massless spin-2 field in flat space. He described his results in his unpublished 1946–47 Bachelor's thesis [Krai 47]. Kraichnan continued to pursue this problem at the Institute for Advanced Study in 1949–50. He recalls that, though he received some

---

[2] See also [BoDe 75], [Dese 87].

encouragement from Bryce DeWitt, very few of his colleagues supported his efforts. This certainly included Einstein himself, who was appalled by an approach to gravitation that rejected Einstein's own hard-won geometrical insights. Kraichnan did not publish any of his results until 1955 [Krai 55, Krai 56], when he had finally found a derivation that he was satisfied with. Unlike Gupta, Kraichnan did not *assume* that gravity couples to the total energy momentum tensor. Rather, like Feynman, he derived this result as a consequence of the consistency of the field equations. It seems likely that Feynman was completely unaware of the work of Gupta and Kraichnan.

We should point out that Feynman's analysis was far from the most general possible (and was considerably less general than Kraichnan's). He assumed a particular form for the matter action (that of a relativistic particle), and further assumed a strictly linear coupling of the spin-2 field to the matter (which would not have been possible for a more general matter action). In particular, we note that all of the physical predictions of the theory are unchanged if a nonlinear local redefinition of the spin-2 field h is performed; we are free to replace $h_{\mu\nu}(x)$ by $\tilde{h}_{\mu\nu}(h(x)) = h_{\mu\nu}(x) + O(h(x)^2)$. Feynman has implicitly removed the freedom to perform such field redefinitions by requiring the coupling to the matter to be linear in h. (Field redefinitions are treated in detail by Boulware and Deser [BoDe 75].) A considerably more general analysis of the consistency condition for the field equation, performed much later by Wald [Wald 86], leads in the end to conclusions similar to those of Kraichnan and Feynman.

A quite different approach to deducing the form of the gravitational interaction was developed by Weinberg [Wein 64a, Wein 64b]. From very reasonable assumptions about the analyticity properties of graviton-graviton scattering amplitudes, Weinberg showed that the theory of an interacting massless spin-2 particle can be Lorentz invariant only if the particle couples to matter (including itself) with a universal strength; in other words, only if the strong principle of equivalence is satisfied. In a sense, Weinberg's argument is the deepest and most powerful of all, since the property that the graviton couples to the energy-momentum tensor is derived from other, quite general, principles. Once the principle of equivalence is established, one can proceed to the construction of Einstein's theory (see [Wein 72]).

Finally, there is the issue of how terms in the Lagrangian involving more that two derivatives of $h_{\mu\nu}$ are to be excluded. Feynman has little to say about this, except for the remark in §6.2 that including only the terms with two or fewer derivatives will lead to the "simplest" theory. (See also §10.3, for a related remark in a slightly different context.) He does not seem to anticipate the modern viewpoint [Wein 79]—that terms with more derivatives are bound to be present, but they have a negligible effect on the theory's predictions when the spacetime curvature is weak. The philosophy underlying this viewpoint is that the Einstein theory's Lagrangian is merely an "effective Lagrangian" that describes the low-energy phenomenology of a more fundamental theory—a theory that might involve new degrees of freedom (superstrings?) at length scales of the order of the Planck length $L_P = (G\hbar/c^3)^{1/2} \simeq 10^{-33}$ cm. In the effective Lagrangian, all terms consistent with general principles are allowed, including terms with arbitrary numbers of derivatives. However, on dimensional grounds, a term with more derivatives will have a coefficient proportional to a higher power of L_P. Thus, in a process involving a characteristic radius of curvature of order L, terms in the Lagrangian with four derivatives have effects that are suppressed compared to the effects of terms with two derivatives—suppressed by a factor of order $(L_P/L)^2$, which is exceedingly small for any reasonable process. We can easily understand, then, why a truncated theory involving only terms with two or fewer derivatives would be in excellent agreement with experiment.

On the other hand, this same reasoning also leads one to expect a "cosmological" term (no derivatives) with a coefficient of order one in units of L_P. That the cosmological constant is in fact extraordinarily small compared to this naive expectation remains one of the great unsolved mysteries of gravitation physics [Wein 89].

Geometry:

After conducting a search for a sensible theory that describes the interactions of a massless spin-2 field in flat space, Feynman does not fail to express delight (in §8.4) that the resulting theory has a geometrical interpretation: "... the fact is that a spin two field has this geometrical interpretation; this is not something readily explainable—it is just marvelous." The development of the theory in Lectures 8–10 makes use of geometrical language, and is much more traditional than the approach of the earlier lectures.

In §9.3, Feynman comments that he knows no geometrical interpretation of the Bianchi identity, and he sketches how one might be found. The geometrical interpretation that he envisions was actually implicit in the 1928 work of the French mathematician Elie Cartan [Cart 28]; however, it was largely unknown to physicists, even professional relativists, in 1962, and it was couched in the language of differential forms, which Feynman did not speak. Cartan's interpretation, that "the boundary of a boundary is zero," was finally excavated from Cartan's ideas by Charles Misner and John Wheeler in 1971, and has since been made widely accessible by them; see, e.g., Chapter 15 of [MTW 73] at the technical level, and Chapter 7 of [Whee 90] at the popular level.

Cosmology:

Some of Feynman's ideas about cosmology have a modern ring. A good example is his attitude toward the origin of matter. The idea of continuous matter creation in the steady state cosmology does not seriously offend him (and he notes in §12.2 that the big bang cosmology has a problem just as bad, to explain where all the matter came from in the *beginning*). In §1.2 and again in §13.3, he emphasizes that the total energy of the universe could really be zero, and that matter creation is possible because the rest energy of the matter is actually canceled by its gravitational potential energy. "It is exciting to think that it costs *nothing* to create a new particle, . . ." This is close to the currently popular view that the universe is a "free lunch," nothing or nearly nothing blown up to cosmological size through the miracle of inflation [Guth 81]. Feynman worries more about the need for baryon number nonconservation, if the universe is to arise from "nothing."

Feynman also expresses a preference for the "critical" value of the density of the universe in §13.1, a prejudice that is widely held today [LiBr 90]. In §13.2, he gives an interesting (and qualitatively correct) argument to support the density being close to critical: he notes that the existence of clusters and superclusters of galaxies implies that "the gravitational energy is of the same order as the kinetic energy of the expansion—this to me suggests that the average density must be very nearly the critical density everywhere." This was a rather novel argument in 1962.

It is evident that, already in the early 60's, Feynman recognized the need for new fundamental principles of physics that would provide a prescription for the initial conditions of the universe. Early in these lectures, in §2.1, he digresses on the foundations of statistical mechanics, so as to express his conviction that the second law of thermodynamics must have a cosmological origin. Note his statement, "The question is how, in quantum mechanics, to describe the idea that the state of the universe in the past was something special." (A similar insight also appears in *The Feynman Lectures on Physics* [Feyn 63a], and *The Character of Physical Law* [Feyn 67], which date from the same period.) Thus, Feynman seems to anticipate the fascination with quantum cosmology that began to grip a segment of the physics community some twenty years later. He also stresses, in §1.4 and §2.1, the inappropriateness of the Copenhagen interpretation of quantum mechanics in a cosmological context.

Superstars:

In 1962–63, when Feynman was delivering his lectures on gravitation, Caltech was awash in new discoveries about "strong radio sources."

For 30 years, astronomers had been puzzling over the nature of these strongest of all radio-emitting objects. In 1951 Walter Baade [Baad 52] had used Caltech's new 200-inch optical telescope on Palomar Mountain to discover that the brightest of the radio sources, Cygnus A, is not (as astronomers had expected) a star in our own galaxy, but rather is associated with a peculiar, somewhat distant galaxy. Two years later,

R. C. Jennison and M. K. Das Gupta [JeDG 53], studying Cygnus A with a new radio interferometer at Jodrell Bank, England, had discovered that most of the radio waves come not from the distant galaxy's interior, but from two giant lobes on opposite sides of the galaxy, about 200,000 light years in size and 200,000 light years apart. Caltech's Owens Valley radio interferometer went into operation in the late 1950s, and by 1962–63, the year of Feynman's gravity lectures, it was being used, together with the Palomar 200 inch, to identify many more double lobed radio sources. Some, like Cyg A, were centered on galaxies; others were centered on star-like point sources of light (which Caltech's Maarten Schmidt on February 5, 1963 would discover to have huge red shifts [Schm 63], and Hong-Yee Chiu later that year would christen *quasars*). In 1962 and early '63, while Caltech's astronomers competed with each other to make new and better observations of these strange objects and interpret their spectra, astrophysicists competed in the construction of models.[3]

One particularly promising model, conceived in summer 1962 by Cambridge's Fred Hoyle and Caltech's William Fowler [HoFo 63], held that the power for each strong radio source comes from a supermassive star in the center of a galaxy. The radio lobes' prodigious energy (estimated by Geoffrey Burbidge as 10^{58} to 10^{60} ergs, i.e., the energy equivalent of 10^4 to 10^6 solar masses) required the powering star to have a mass in the range $\sim 10^6$ to 10^9 suns. By comparison with the ~ 100 solar mass upper limit for normal stars, these Hoyle-Fowler objects were indeed "supermassive." *Superstars* they came to be called in some circles.

Sometime in early 1963 (probably February or March), Fred Hoyle gave a SINS[4] Seminar, in Caltech's Kellogg Radiation Laboratory, on the superstar model for strong radio sources. During the question period, Richard Feynman objected that general relativistic effects would make all superstars unstable—at least if they were nonrotating. They would collapse to form what we now call black holes.

Hoyle and Fowler were dubious, but within a few months, they and independently Icko Iben [Iben 63] (a Senior Research Fellow in Fowler's Kellogg Lab) had verified that Feynman was probably right, and S. Chandrasekhar [Chan 64] at the University of Chicago had independently discovered the general relativistic instability and analyzed it definitively.

To Hoyle and Fowler, Feynman's remark was a "bolt out of the blue," completely unanticipated and with no apparent basis except Feynman's amazing physical intuition. Fowler was so impressed, that he described the seminar and Feynman's insight to many colleagues around the world, adding one more (true) tale to the Feynman legend.

Actually, Feynman's intuition did not come effortlessly. Here, as else-

---

[3] For further historical detail see, e.g., Chapter 9 of [Thor 94] and references therein.

[4] "Stellar Interiors and Nucleosynthesis," a seminar series that Fowler organized and ran.

where, it was based in large measure on detailed calculations driven by Feynman's curiosity. And in this case, in contrast with others, Feynman has left us a detailed snapshot of one moment in the struggle by which he made his discovery: Lecture 14 of this volume.

We have pieced together the circumstances surrounding Lecture 14, largely from Icko Iben's January 1963 notes and memory, plus a ca. 1971 conversation between Feynman and Thorne, and inputs from James Bardeen, Steven Frautschi, James Hartle, and William Fowler:

Sometime in late 1962 or early January 1963, it must have occurred to Feynman that the Hoyle-Fowler superstars might be strongly influenced by general relativistic forces. According to Iben's notes, Feynman came to his office in Kellogg Lab sometime before January 18, raised the issue of how general relativity influences superstars, showed Iben the general relativistic equations that govern a superstar's structure, which Feynman had worked out for himself from first principles, and asked how astrophysicists like Iben go about building Newtonian stellar models from the analogous Newtonian equations. After their discussion, Feynman departed, and then returned a few days later, sometime during the week of January 21–25. "Feynman flabbergasted me," Iben recalls, "by coming in and telling me that he had [already] solved the ... equations. He told me that he was doing some consulting for a computer firm and solved the equations in real time on what must have been that generation's version of a workstation."

On Monday January 28, having had only a few days to think about his numerical solutions (and presumably having spent much of that time in other ways, since he also had to prepare a sophomore physics lecture for that same Monday), Feynman delivered Lecture 14 of this volume. (Note that this was just eight days before Maarten Schmidt's discovery of the quasar redshifts.)

Lecture 14 came in the midst of Feynman's struggle to figure out how superstars should behave—and it came *before* he realized that general relativity destabilizes them. As a result, the interpretation portions of Lecture 14 (§14.3 and §14.4) are largely wrong—but wrong in interesting ways that display Feynman's intuitive approach to problem solving.

Feynman never reviewed the Morinigo-Wagner written version of Lecture 14; and in 1971, when he approved Lectures 12–16 for distribution, he presumably had forgotten that the interpretation part of Lecture 14 was just a progress report on ruminations that had not yet borne fruit.

Feynman begins his Lecture 14 by introducing a model for a superstar "which is very simple, yet may possess a great many of the attributes of the real things. After we understand how to go about solving the simple problem, we may worry about refinements in the model." (The refinements—the influence of electron-positron pairs, neutrino emission, nuclear burning, rotation, and instabilities—would be added later in 1963–64 by Iben [Iben 63], Curtis Michael [Mich 63], Fowler [Fowl 64], and Bardeen [Bard 65], with major advice and input from Feynman.)

Because Feynman's goal is to learn about relativity's effects, his superstar model is fully general relativistic—by contrast with the previous Fowler-Hoyle models, which were Newtonian. On the other hand, where Fowler and Hoyle had included the contributions of both gas and radiation to the star's pressure p and internal energy density ϵ, Feynman simplifies by ignoring p_{gas} and ϵ_{gas}. This seems reasonable, since Feynman's principal focus was on a superstar with mass $M = 10^9 M_{sun}$, and Hoyle and Fowler had shown that in the Newtonian limit superstars are strongly radiation dominated, with

$$\beta \equiv \frac{p_{gas}}{p_{radiation}} = \frac{2\epsilon_{gas}}{\epsilon_{radiation}} = 8.6 \left(\frac{M_{sun}}{M}\right)^{1/2} \simeq 3 \times 10^{-4} \left(\frac{10^9 M_{sun}}{M}\right)^{1/2}.$$

(F.2)

(Here, for simplicity, the gas is assumed to be pure hydrogen.) Because these stars are deeply convective, their entropy per nucleon is, independent of radius, which means that β, which is 8 × (Boltzmann's constant)/(entropy per nucleon), is also radius-independent; and this, as Feynman was aware, remains true in the fully relativistic regime, though Eq. (F.2) is changed there by a factor of order unity.

Having approximated away p_{gas} and ϵ_{gas}, Feynman proceeds, in §14.1 and §14.2 of Lecture 14, to construct the superstar's general relativistic equations of structure, he reports that he has integrated them numerically, and he presents his results in Table 14.1. This table is to be interpreted with the aid of equations (14.2.1), in which Feynman's parameter τ is

$$\tau = \frac{4/3}{\text{entropy per nucleon}} = \left(\frac{\text{nucleon rest mass}}{\text{Boltzmann's constant}}\right) \frac{\beta}{6} \simeq 1800\beta,$$

(F.3)

since Feynman uses units with the nucleon rest mass and 10^9 °K set to unity.

In discussing Feynman's models and his (mis)interpretation of them, Figure F.1 will be useful. This figure shows some features of the family of superstar models that Feynman constructed (thick curve), together with their extension into the ultrarelativistic regime (upper thin curve) and into the nearly Newtonian regime (lower thin curve)—extensions that would be computed later by Iben [Iben 63], Fowler [Fowl 64], Bardeen [Bard 65], and Tooper [Toop 66]. Plotted vertically is the negative of the star's gravitational binding energy; plotted horizontally is the star's radius. In the nearly Newtonian regime (shaded region of the binding energy curve), which Feynman does not explore, p_{gas} and ϵ_{gas} cannot be neglected, and the binding energy is given (as Fowler [Fowl 64] would

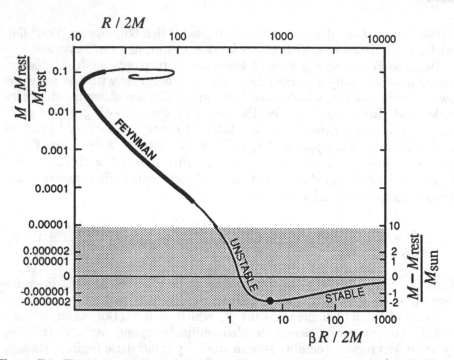

Figure F.1 The binding energies of superstars made of hydrogen. Plotted vertically on the left is the negative of the star's fractional binding energy, i.e., $(M - M_{\text{rest}})/M_{\text{rest}}$, where M is the star's total mass and M_{rest} is the total rest mass of all its nucleons; plotted horizontally on the top is the star's radius R in units of the Schwarzschild radius $2M$ of a black hole of the same mass. The left and top scales are valid in the white region for superstars of any mass, but in the (nearly Newtonian) shaded region only for $M = 10^6 M_{\text{sun}}$. Plotted vertically on the right is the negative of the star's binding energy in units of the sun's mass M_{sun}; on the bottom is $R/2M$ multiplied by the ratio, β, of gas pressure to total pressure. The right and bottom scales are valid for superstars of any mass in the nearly Newtonian shaded region, but they fail in the fully relativistic white region. The vertical scale is arctangent; i.e., it is nearly linear for $|(M - M_{\text{rest}})/M_{\text{sun}}| \lesssim 1$ and logarithmic for $|(M - M_{\text{rest}})/M_{\text{sun}}| \gtrsim 1$. The thick part of the curve is from Feynman's calculations in Lecture 14; the thin parts are due to Fowler [Fowl 64], Iben [Iben 63], Bardeen [Bard 65], and Tooper [Toop 66].

show in response to Feynman's "bolt out of the blue") by the following delicate balance between gas effects (first term) and general relativistic effects (second term):

$$\frac{M - M_{\text{rest}}}{M_{\text{rest}}} \simeq -\frac{3\beta}{8}\left(\frac{2M}{R}\right) + 1.3\left(\frac{2M}{R}\right)^2. \qquad \text{(F.4)}$$

Here $2M$ is the Schwarzschild radius of a black hole with the same mass M as the superstar.

In interpreting his models (§14.3), Feynman begins by asking about the evolution of a superstar that contains a fixed number of nucleons (i.e., a fixed nucleon rest mass M_{rest}), and that gradually radiates away thermal energy, thereby reducing its total mass M and making itself more tightly bound. He finds a strange evolution: As the star radiates, its radius increases (downward and rightward motion on the thick curve of Fig. F.1), and its central temperature goes down. This is opposite to the behavior of most stars, which shrink and heat up as they radiate, if they are not burning fuel. (If, instead of dealing with the fully relativistic region to the left of the binding curve's minimum, Feynman had kept gas effects and computed the nearly Newtonian region to the right, he would have found the opposite behavior: the superstar would shrink and heat up as it radiates.)

Feynman then asks whether his superstar models are stable. "The stability of our star has not been studied [quantitatively]," he emphasizes, and he then goes on to expose his initial ruminations on the issue:

"[Models which have] the same number of nucleons and the same τ [same entropy] may be compared as to radii and central temperature. The fact that there is apparently a minimum in the radius [left-most bend in Fig. F.1] ... is very suggestive; the star may have a stable position." Feynman here is groping toward a method of analyzing stability that would be perfected a year or so later by James Bardeen, when Bardeen had become Feynman's Ph.D. thesis student. Bardeen's perfected version of this argument [Bard 65, BTM 66] shows that, as one moves along the binding-energy curve, restricting oneself to stars of fixed M_{rest}, the entropy changes from one model to another except in the vicinity of each minimum or maximum of the binding, where it is stationary. This means that the star possesses a zero-frequency mode of deformation at each minimum or maximum—a mode that carries it from one equilibrium model to another with the same entropy, binding energy, and rest mass. This in turn means that one mode of radial oscillation changes stability at each extremum of the binding. By examining the forms that the modes' eigenfunctions must take, Bardeen deduces that, if the binding curve bends clockwise as it moves through an extremum, then the mode is becoming unstable; and if counterclockwise, then the mode is becoming stable. (This statement holds regardless of the direction one traces the curve.) Bardeen's analysis, applied to Fig. F.1, shows that the nearly Newtonian models in the lower right (which shrink as they radiate) are stable, and they must lose stability and collapse to form a black hole when they reach the minimum of the binding curve; the models beyond the minimum (including all of Feynman's models) possess one unstable mode of radial pulsation; the models beyond the first peak in the binding curve (upper left part of Fig. F.1) possess two unstable modes; etc.

Feynman, of course, was not aware of this on January 28, 1963; so he goes on in Lecture 14 to seek insight into his models' stability by other methods. He imagines taking one of his superstar models with baryonic rest mass M_{rest}, and breaking it up into two superstars, each with rest mass $M_{rest}/2$, while holding the entropy per nucleon fixed. "Do we get work out of the process, or did we have to do work to get [the star] broken up?" From his Table 1 and equations (14.2.1) he deduces that "the two objects ... would be more massive; work is required to break up the system. This suggests that the star might not throw off material, but keep together in one lump," i.e., the star might be stable.

At first sight this seems a compelling argument. However, it actually is specious (as Feynman presumably realized sometime between this lecture and Hoyle's seminar). The two new stars that Feynman makes by breaking up his original star are farther up the thick binding-energy curve of Fig. F.1, i.e., they are more relativistic than the original star. However, there are also two stellar models, with the same rest masses and entropy per nucleon, on the stable, nearly Newtonian branch of the binding curve in the lower right corner of Fig. F.1. If the original star were broken up into those two stars, energy would be released, which suggests correctly that the original star is actually unstable. Feynman missed this point because neither he, nor anyone else on January 28 1963, knew the form of the binding energy curve in the shaded region. It is reasonable to surmise, however, that he guessed enough about it before Hoyle's seminar to recognize his error.

Having misdiagnosed his stars' stability, Feynman goes on in §14.4 to propose directions for future superstar research. He begins by proposing a variational principle by which one might construct equilibrium models of fully relativistic, isentropic superstars: "compute the configuration of least mass starting with a fixed number of nucleons" (and fixed entropy per nucleon). Two years later John Cocke [Cock 65], working in Paris and presumably unaware of Feynman's proposal, would develop a detailed variational principle equivalent to Feynman's (holding the mass and nucleon number fixed and maximizing the entropy), and would use it to construct relativistic stellar models.

Feynman continues in §14.4 with, "After we have investigated static solutions, we may turn our attention to the full dynamical problem. The differential equations are horrifying." Feynman, in fact, had worked out some of the equations for himself. They would later be derived independently and solved numerically by Mikhail Podurets [Podu 64] in the USSR and Michael May and Richard White [MaWh 66] in the United States, using descendants of computer codes that were developed for the design of thermonuclear weapons. The result is well known: stars that experience the Feynman/Chandrasekhar relativistic instability implode to form black holes.

For about 10 years after Feynman's Lecture 14, rapidly rotating superstars remained a strong competitor in the marketplace of powerhouse models for quasars and strong radio sources. But gradually, in the 1970s, models based on rapidly spinning, supermassive black holes gained ascendancy; and today superstars are generally regarded as fascinating but transient objects along a galaxy core's route toward formation of the supermassive black hole that will later come to dominate it; cf. [Thor 94].

Black holes:

The concept of a black hole was just beginning to emerge in the early 60's, and Feynman's views may have been slightly behind the times. Thus, the most seriously out-of-date lectures are probably 11 and 15, which deal with the Schwarzschild solution and its implications.

In one sense, what we now call a black hole already became known in 1916, when Karl Schwarzschild discovered his solution to the Einstein field equation [Schw 16]. But for decades most physicists stubbornly resisted the outrageous implications of Schwarzschild's solution. (This included Einstein himself, who wrote a regrettable paper in 1939 arguing that black holes cannot exist [Eins 39].) Even the beautiful and definitive analysis (also in 1939) of gravitational collapse by Oppenheimer and Snyder [OpSn 39] had surprisingly little impact for many years. Oppenheimer and Snyder studied the collapse of a spherically symmetric pressureless "star" of uniform density, and noted that the star's implosion, as seen by a stationary observer who remains outside, would slow and ultimately freeze as the star's surface approached the critical Schwarzschild circumference. Yet they also clearly explained that no such freezing of the implosion would be seen by observers riding in with the collapsing matter—these observers would cross inside the critical circumference in a finite proper time, and thereafter would be unable to send a light signal to observers on the outside. This extreme difference between the descriptions in the two reference frames proved exceptionally difficult to grasp. The two descriptions were not clearly reconciled until 1958, when David Finkelstein [Fink 58] analyzed the Schwarzschild solution using a coordinate system that made it easy to visualize simultaneously the worldlines of dust particles that fall inward through the critical circumference and the worldlines of outgoing photons that freeze there. This analysis revealed the unusual "causal structure" of the Schwarzschild spacetime—nothing inside the "horizon" can avoid being drawn toward spheres of smaller and smaller area. The emerging picture indicated (to some), that once the star fell through its critical circumference, its compression to form a spacetime singularity was inevitable. That this, indeed, is true, independent of any idealizing assumptions such as spherical symmetry and zero pressure, would be proved by Roger Penrose in 1964 [Penr 65].

Thus, the timing of Feynman's lectures on gravitation is unfortunate. A "golden age" of black hole research was about to begin, which would produce many remarkable insights over the next decade. These insights, which could not have been anticipated in 1962–63, would completely transform the study of general relativity, and help usher in the new discipline of relativistic astrophysics.

Feynman's 1962–63 view of the Schwarzschild solution was heavily influenced by John Wheeler. Wheeler had felt for many years that the conclusions of Oppenheimer and Snyder could not be trusted; he thought them physically unreasonable. As late as 1958, he advocated that, if a more realistic equation of state were used in the analysis of gravitational collapse, qualitatively different results would be obtained [HWWh 58]. (This view would become less tenable as the causal structure of the black hole geometry became properly understood.) Gradually, however, Wheeler came to accept the inevitability of gravitational collapse to form a black hole, in agreement with the conclusions of Oppenheimer-Snyder. (This shift of viewpoint was facilitated by the insights of Martin Kruskal [Krus 60], who, independently of Finkelstein, had also clarified the black hole causal structure; in fact, Kruskal's influential paper was largely written by Wheeler, though the insights and calculations were Kruskal's.) But during the years that he remained skeptical, Wheeler reacted in a characteristic way—he rarely mentioned the Oppenheimer-Snyder results in his published papers. It is revealing that in §11.6 Feynman remarks that it would be interesting to study the collapse of dust. He seems unaware that Oppenheimer and Snyder had studied dust collapse in detail 23 years earlier! In §15.1 he speculates, based on the (incorrect!) ruminations of Lecture 14, that a star composed of "real matter" cannot collapse inside its critical circumference.

Feynman makes several references to the "geometrodynamics" program that Wheeler had been pushing since the mid 50's, and was still pushing (if less vigorously) in 1962; cf. [Whee 62]. Wheeler and coworkers hoped to interpret elementary particles as geometric entities arising from (quantum versions of) classical solutions to the matter-free gravitational field equations. Wheeler was especially attracted to the concept of "charge without charge;" he noted that if electrical lines of force were trapped by the nontrivial topology of a "wormhole" in space, then each wormhole mouth would appear to be a pointlike charged object to an observer whose resolution is insufficient to perceive the tiny mouth [MiWh 57]. Wheeler emphasized that the Schwarzschild solution possesses spatial slices in which two asymptotically flat regions are connected by a narrow neck, and so provides a model of the wormhole geometry that he envisioned.

Feynman is clearly enamored of the wormhole concept, and he describes these ideas briefly in §11.5, and then again in §15.1 and §15.3. Note that Feynman calls a star confined inside its gravitational radius a "wormhole;" the term "black hole" would not be coined (by Wheeler) until 1967. For what we now call the "horizon" of a black hole, Feynman uses the older term "Schwarzschild singularity." This is an especially unfortunate locution, as it risks confusion with the actual singularity, the region of infinite spacetime curvature at the hole's center. Feynman never explicitly discusses this genuine singularity.

By 1962, the causal structure of the Schwarzschild solution was fairly well understood. It is well explained by Fuller and Wheeler [FuWh 62], a paper that Feynman mentions, and on which §15.1 is based. (This paper, one of the very few references cited in Feynman's lectures, employs Kruskal's coordinates to construct the complete analytically extended Schwarzschild geometry, and presents a "Kruskal diagram" that succinctly exhibits the properties of the timelike and null geodesics.) Feynman quotes the main conclusion: the Schwarzschild solution is not really a wormhole of the sort that Wheeler is interested in, because the wormhole throat is actually dynamical, and pinches off before any particle can traverse it. However, the Fuller-Wheeler paper does not mention any of the broader implications of this causal structure for the problem of gravitational collapse, and Feynman gives no indication of having appreciated those implications.

One can also see from Feynman's comments in §15.2 and §15.3 that he did not understand the causal structure of the ("Reissner-Nordström") charged black hole solution, which had been worked out in 1960 by Graves and Brill [GrBr 60]. Note the remark "... it is not inconceivable that it might turn out that a reflected particle comes out earlier than it went in!" In fact, on the analytically extended geometry, the geodesics pass into a "new universe" in finite proper time, rather than reemerging from the black hole (see, e.g., [HaEl 73]). However, the interior of this solution is known to be unstable to generic perturbations [ChHa 82]; for the "realistic" case of a charged black hole formed in gravitational collapse, the situation is qualitatively different, and still not fully understood—though it seems highly likely that the hole's core is so singular that nothing can pass through it to a "new universe," at least within the realm of classical general relativity [BBIP 91].

Gravitational waves:

As late as 1957, at the Chapel Hill conference, it was still possible to have a serious discussion about whether Einstein's theory really predicted the existence of gravitational radiation [DeWi 57]. This confusion arose in large measure because it is a rather subtle matter to define rigorously the energy transmitted by a gravitational wave—the trouble is that the gravitational energy cannot be expressed as the integral of a locally measurable density.

At Chapel Hill, Feynman addressed this issue in a pragmatic way, describing how a gravitational wave antenna could in principle be designed that would absorb the energy "carried" by the wave [DeWi 57, Feyn 57].

In Lecture 16, he is clearly leading up to a description of a variant of this device, when the notes abruptly end: "We shall therefore show that they can indeed heat up a wall, so there is no question as to their energy content." A variant of Feynman's antenna was published by Bondi [Bond 57] shortly after Chapel Hill (ironically, as Bondi had once been skeptical about the reality of gravitational waves), but Feynman never published anything about it. The best surviving description of this work is in a letter to Victor Weisskopf completed in February, 1961 [Feyn 61]. This letter contains some of the same material as Lecture 16, but then goes a bit further, and derives the formula for the power radiated in the quadrupole approximation (a result also quoted at Chapel Hill). Then the letter describes Feynman's gravitational wave detector: It is simply two beads sliding freely (but with a small amount of friction) on a rigid rod. As the wave passes over the rod, atomic forces hold the length of the rod fixed, but the proper distance between the two beads oscillates. Thus, the beads rub against the rod, dissipating heat. (Feynman included the letter to Weisskopf in the material that he distributed to the the students taking the course.)

However controversial they may have seemed to some of the participants at the Chapel Hill meeting, Feynman's conclusions about gravitational waves were hardly new. Indeed, a classic textbook by Landau and Lifshitz, which was completed in Russian in 1939 and appeared in English translation in 1951 [LaLi 51], contains several sections devoted to the theory of gravitational waves. Their account is clear and correct, if characteristically terse. In the letter to Weisskopf, Feynman recalls the 1957 conference and comments, "I was surprised to find a whole day at the conference devoted to this question, and that 'experts' were confused. That is what comes from looking for conserved energy tensors, etc. instead of asking 'can the waves do work?' "

Indeed, in spite of his deep respect for John Wheeler, Feynman felt an undisguised contempt for much of the relativity community in the late 50's and early 60's. This is perhaps expressed most bluntly in a letter to his wife Gweneth that he wrote from the Warsaw conference in 1962 [Feyn 88]:

> I am not getting anything out of the meeting. I am learning nothing. Because there are no experiments this field is not an active one, so few of the best men are doing work in it. The result is that there are hosts of dopes here and it is not good for my blood pressure: such inane things are said and seriously discussed that I get into arguments outside the formal sessions (say, at lunch) whenever anyone asks me a question or starts to tell me about his "work." The "work" is always: (1) completely un-understandable, (2) vague and indefinite, (3) something correct that is obvious and self-evident, but worked out by a long and difficult analysis, and presented as an important discovery, or (4) a claim based on the stupidity of the author that some obvious and correct

fact, accepted and checked for years, is, in fact, false (these are the worst: no argument will convince the idiot), (5) an attempt to do something probably impossible, but certainly of no utility, which, it is finally revealed at the end, fails, or (6) just plain wrong. There is a great deal of "activity in the field" these days, but this "activity" is mainly in showing that the previous "activity" of somebody else resulted in an error or in nothing useful or in something promising. It is like a lot of worms trying to get out of a bottle by crawling all over each other. It is not that the subject is hard; it is that the good men are occupied elsewhere. Remind me not to come to any more gravity conferences!

So extreme an assessment could not have been completely justified even in 1962, nor does it seem likely that the letter reflected Feynman's true feelings with 100 per cent accuracy. Bryce DeWitt, who attended both the Chapel Hill and Warsaw conferences, offers this comment [DeWi 94]:

I can surely sympathize with Feynman's reaction to the Warsaw conference because I had similar feelings. (I have a vivid memory of him venting his frustrations there by giving Ivanenko one of the most thorough tongue lashings I have ever heard.) But those who publish his private letter without giving the whole picture do a disservice to historical truth. Although he felt that some of the discussion at the Chapel Hill conference was nonsensical (as did I), I think he had a reasonably good time there. I remember him being very interested when I showed that his path integral for a curved configuration space leads to a Schroödinger equation with a Ricci scalar term in it. The people at that conference (such as Bondi, Hoyle, Sciama, Møller, Rosenfeld, Wheeler) were not stupid and talked with him intelligently. (I had chosen the participants myself—it was a closed conference.) Feynman's experience at Chapel Hill surely had something to do with his willingness to accept the invitation to Warsaw (which was an open conference). Even at Warsaw he and I had discussions outside the formal session that I try not to believe he could honestly have put into one of the six categories in his letter.

However apt Feynman's comments may or may not have been in 1962, they would soon cease to be so. The "golden age" of black hole research was just beginning to dawn.

Philosophy:

A striking feature of these lectures is that Feynman is frequently drawn to philosophical issues. (He often showed disdain for philosophers of science and for the word "philosophical," which he liked to mockingly pronounce "philo-ZAW-phical;" nevertheless, he is revered, at least by physicists, for his philosophical insights.) For example, he ruminates in §1.4 on whether quantum mechanics need actually apply to macroscopic objects. (The argument he sketches there to support the claim that quantization of gravity is really necessary was presented at the 1957 Chapel Hill conference, where it provoked a great deal of discussion.) Another example is his preoccupation with Mach's Principle. Mach's idea—that inertia arises from the interactions of a body with distant bodies—bears a vague resemblance to the interpretation of electrodynamics proposed by Feynman and Wheeler when Feynman was in graduate school [WhFe 45, WhFe 49]: that the radiation reaction force on an accelerated charge arises from interactions with distant charges, rather than with the local electromagnetic field. So perhaps it should not be surprising that in §5.3 and §5.4 Feynman seems sympathetic to Mach's views. He gropes for a quantum mechanical formulation of Mach's Principle in §5.4, and revisits it in a cosmological context in §13.4. Feynman's reluctance in §9.4 and §15.3 to accept the idea of curvature without a matter source also smacks of Mach.

Philosophical reflections come to the fore in a number of brief asides. In §8.3 Feynman assesses the meaning of the statements that space is "really" curved or flat. In §7.1 he explains why Newton's second law is not a mere tautology (a definition of "force"). He takes some slaps at rigor in §10.1 ("it is the facts that matter, and not the proofs") and in §13.3 ("there is no way of showing mathematically that a physical conclusion is wrong or inconsistent"). And in §13.4 he questions the notion that simplicity should be a guiding principle in the search for the truth about Nature: "...the simplest solution by far would be nothing, that there should be *nothing* at all in the universe. Nature is more inventive than that, so I refuse to go along thinking that it always has to be simple."

It is also revealing when Feynman bullheadedly pursues an unpromising idea in §2.3 and §2.4—that gravitation is due to neutrino exchange. This is an object lesson on how Feynman believes a scientist should react to a new experimental phenomenon: one should always look carefully for an explanation in terms of known principles before indulging in speculations about new laws. Yet, at the same time, Feynman emphasizes over and over again the importance of retaining skepticism about accepted ideas, and keeping an open mind about ideas that appear flaky. Quantum mechanics might fail (§1.4 and §2.1), the universe might not be homogeneous on large scales (§12.2 and §13.2), the steady state cosmology could turn out to be right (§13.3), Wheeler's intuition concerning wormholes might be vindicated (§15.3), etc.

One-loop quantum gravity:

Feynman's investigations of quantum gravity eventually led him to a seminal discovery (which, however, is not described in this book, except for a brief mention in §16.2). This is his discovery that a "ghost" field must be introduced into the covariantly quantized theory to maintain unitarity at one-loop order of perturbation theory. The timing of this discovery can be dated fairly precisely. Feynman reported the result in a talk at a conference in Warsaw in July, 1962 [Feyn 63b], and commented that the problem of unitarity at one-loop order had only been "completely straightened out a week before I came here." Thus, he had it all worked out before he gave these lectures.

By computing one-loop amplitudes using the naive covariant Feynman rules, Feynman had found that the contributions of the unphysical polarization states of the graviton fail to cancel completely, resulting in a violation of unitarity. For awhile, he was unable to resolve this puzzle. Then Murray Gell-Mann suggested to Feynman that he try to analyze the simpler case of massless Yang-Mills theory. (Gell-Mann recalls making this suggestion in 1960 [Gell 89].) Feynman found that he could fix up the problem in Yang-Mills at one loop, and then could use a similar method for gravity. In his Warsaw talk, Feynman reports that he has solved the unitarity problem at one loop order, but that he is now stuck again—he doesn't know how to generalize the method to two or more loops. Yet he protests that, "I've only had a week, gentlemen." He never solved this problem.

There is an interesting exchange in the question period following Feynman's talk at Warsaw [Feyn 63b], in which Bryce DeWitt presses Feynman for more detail about how unitarity is achieved at one-loop order, and Feynman resists. But DeWitt is persistent, and Feynman finally relents, and offers a rather lengthy explanation, prefaced by the comment, "Now I will show you that I too can write equations that nobody can understand." This is amusing because it was eventually DeWitt [DeWi 67a, DeWi 67b] (and also Faddeev and Popov [FaPo 67], independently) who worked out how to generalize the covariant quantization of Yang-Mills theory and gravitation to arbitrary loop order. It is worth noting that Feynman's own path integral techniques were crucial to the general formulation, and that the more complete results of DeWitt and Faddeev-Popov were clearly inspired by Feynman's one-loop construction.

Feynman's lectures late in the gravitation course (the ones that are not reproduced in this book) concerned loop corrections in quantum gravity and Yang-Mills theory. He described the one-loop results that had been reported in Warsaw, and his attempts to extend the results to higher order. By all accounts, these lectures were complicated and difficult to follow, and sometimes flavored by a palpable sense of frustration. We presume that Feynman felt embarrassment at having failed to find a satisfactory formulation of the perturbation expansion, and that this is why he never authorized the distribution of the notes for these lectures.

Feynman wrote up a detailed account of his results only much later [Feyn 72], in two articles for a volume in honor of John Wheeler's 60th birthday. These articles would never have been written had it not been for persistent badgering by one of us (Thorne). By mutual agreement, Thorne telephoned Feynman at home once a week, at a prearranged time, to remind him to work on his contributions to the Wheeler Festschrift. This continued until the articles were finally completed. It was clear that Feynman returned to quantum gravity only with some pain and regret.

One of Feynman's goals was to settle the issue of the renormalizability of the theory. At the Warsaw meeting, he says that he's not sure if it is renormalizable, but in §16.2 he puts it more strongly: "I suspect that the theory is not renormalizable." Even this statement sounds surprisingly cautious to us today. In any case, Feynman was always reluctant to use renormalizability as a criterion for judging a theory, and he professes in §16.2 not to know whether nonrenormalizability is "a truly significant objection."

Conclusion:

Any book about gravitation prepared over 30 years ago is inevitably out of date today, at least in some respects. This book is certainly no exception. Moreover, we believe that the lectures did not meet Feynman's own expectations, even at the time they were given. He had hoped that teaching this course would help bring his work on quantum gravity to a coherent conclusion, but it did not do so. Toward the end of the year, it was evident to the students that Feynman felt discouraged and frustrated. Thus, in accord with Feynman's own wishes, the lectures (17 through 27) that focused on issues in quantum gravity are not included in this book.

Still, we feel that there is much in this volume's Lectures 1–16 that will be valued by physicists, students, historians, and admirers of Feynman. Moreover, the lectures are fun. Many passages offer a glimpse of a great mind approaching deep and challenging questions from an original perspective. Feynman thought long and hard about gravitation for several years, yet he published remarkably little on the subject. These characteristically clear and clever lectures are an addition to his published oeuvre that should surely be welcomed.

We are grateful to Jim Bardeen, Stanley Deser, Bryce DeWitt, Willy Fowler, Steve Frautschi, Judy Goodstein, Jim Hartle, Icko Iben, Bob Kraichnan, Charles Misner, Fernando Morinigo, Jim Peebles, Allan Sandage, and Bill Wagner for helpful advice that contributed to the preparation of this foreword.

Caltech, May 1995

Quantum Gravity

Brian Hatfield

Feynman gave these lectures on gravitation at Caltech in 1962-63, near the end of the period he often referred to as his "gravity phase." His main motivation for conducting research in gravity at that time was his interest in quantum gravity. He told me in 1980, that he had thought during the fifties that the consequences of quantum gravity might be a "piece of cake" to work out. After all, gravity is really *weak*. Following the spectacular success of perturbative Quantum Electrodynamics, he figured that there would be essentially no need to work out anything beyond first order. Of course, he expected that there might be difficulties in defining a consistent quantum theory (for example, the dimension of the gravitational constant is an obstacle to renormalization). However, his idea was *not* to attempt a complete and consistent quantization, then work out the results, but instead to go in the other direction. That is, compute the perturbative amplitudes for specific processes, such as Compton scattering by a graviton, and then tackle any interesting difficulties that might arise one-by-one. By definition, "interesting" difficulties would be new and unfamiliar problems associated with gravity that had not popped up before in quantum field theory. Thus, initially Feynman ignored the ultraviolet divergences and questions of renormalizability, and only struggled with them later. Ultimately, the lack of a renormalizable formulation led to the abandonment of perturbative quantum gravity. But, characteristically, Feynman's "plan of attack" led him to an important discovery in field theory, namely the need for covariant ghosts to maintain unitarity at one-loop (see the discussion on page xxviii in the Foreword).

These lectures occurred over 30 years ago. We can look back at some of Feynman's analysis in quantum gravity and look at directions that research has taken.

The Relation of Geometry and Quantum Field Theory:

The standard and historical approach to classical gravitation is to start with the Principle of Equivalence and to develop the geometrical viewpoint. Feynman was proud of the fact that he seldom followed the standard approach. On the corner of one blackboard in his office he wrote "What I cannot create, I do not understand." This expression remained virtually untouched on that corner of the blackboard for over 7 years. I first saw it in the fall of 1980, and it was still there in February 1988 (see [Feyn 89]). Thus it is no surprise that Feynman would recreate general relativity from a nongeometrical viewpoint. The practical side of this approach is that one does not first have to learn some "fancy-schmanzy" (as he liked to call it) differential geometry in order to study gravitational physics. (Instead, one would *just* need to learn some quantum field theory.) However, when the ultimate goal is to quantize gravity, Feynman felt that the geometrical interpretation just stood in the way. From the field theoretic viewpoint, one could avoid actually defining—up front—the physical meaning of quantum geometry, fluctuating topology, space-time foam, etc., and instead look for the geometrical meaning after quantization. (See, for example, Sachs question and Feynman's answer in [Feyn 63b]). Feynman certainly felt that the geometrical interpretation is "marvelous" (§8.4) but the fact that a massless spin-2 field can be interpreted as a metric was simply a "coincidence" that "might be understood as representing some kind of gauge invariance."

Presently we have a geometrical interpretation of classical gauge theories such as electrodynamics and Yang-Mills (see, for example, [Yang 77]). The vector potentials A_μ^a are connection coefficients on a principal fiber bundle where the structure group is the gauge group ($U(1)$ for electromagnetism, $SU(2)$ for Yang-Mills, and $SU(3)$ for classical chromodynamics). The field strengths $F_{\mu\nu}$ (i.e., the electric and magnetic fields in electrodynamics) are the curvatures associated with the connections (the potentials). The charged matter that the fields couple to are associated vector bundles (see, for example, [DrMa 77]). Hence, it appears that Feynman's intuition about the connection between geometry and gauge invariance is correct. From his path integral viewpoint, Quantum Electrodynamics and Quantum Chromodynamics amount to integrals over the space of connections on principal fiber bundles. While it can be argued that the geometric interpretation of gauge fields has not helped us solve QED or QCD (i.e., adequately compute or approximate those integrals), it has certainly led to many useful insights into the topological aspects of these theories (e.g., the Gribov ambiguity, instantons, the vacuum angle and topologically inequivalent vacuums), and to the construction of new Yang-Mills type gauge theories with topological masses.

Spin of the Graviton and Antigravity:

One benefit of the field theoretic development of the theory of gravitation is the fact that the (on-shell) graviton is massless and has spin 2 is made explicit without having to start with a fully consistent, generally covariant theory, i.e., without invoking the Principle of General Covariance. This is like building a gravity theory from the bottom up, instead of from the top down using the full geometric apparatus. This development begins in §2.3 of the lectures, and is carried on in §3.1-§3.4. A summary of the argument follows:

In quantum field theory of point particles, the force between two particles is mediated by the exchange of virtual (or off-shell) particles. Associated with each force is a charge. Charged particles feel the force by coupling to or interacting with the particles that carry the force. The most familiar example is electrodynamics. The particles that feel the force carry electric charge. The electromagnetic force is mediated by the exchange of spin-1 photons. The photons themselves are uncharged and therefore do not directly couple to each other. The resulting field equations are linear. In QCD, a theory of the strong force built from a Yang-Mills gauge theory (the strong force is responsible for holding together nucleons and thereby the nucleus), the charge is called color. The fundamental particles that feel the strong force are colored quarks, and the particles that carry the force are called gluons. The gluons themselves are color charged, hence, unlike the photon, they can directly interact with each other, and the resulting field equations are nonlinear. The charge associated with gravity is mass, which we expect from special relativity to be equivalent to energy. Since everything we know about has energy, it appears that gravity should couple to everything. The particle that mediates the gravitational force is called the graviton. Since a graviton has energy, gravitons can directly interact with each other.

If a field theory is to describe gravity, then it must reproduce Newton's Law of Gravitation in the appropriate static nonrelativistic limit, i.e., we must recover

$$F = -\frac{Gm_1m_2}{r^2} \qquad (Q.1)$$

by the exchange of a graviton between particles 1 and 2 separated by r in the appropriate limit. As is well known, the gravitational force is long-ranged (force proportional to $1/r^2$, potential proportional to $1/r$), hence the on-shell or free isolated graviton must be massless, just as in the case of the photon. However, unlike the electromagnetic case, like charges in gravity attract.

In order to produce a *static* force and not just scattering, the emission or absorption of a single graviton by either particle must leave both particles in the same internal state. This rules out the possibility that the graviton carries half-integer spin (for example, related to the fact that it takes a rotation of 720° to return a spin-1/2 wavefunction back to itself). Therefore the graviton must have integer spin. Next, to decide which integer spins are possible, we examine the two cases where particle 2 is identical to particle 1 and where particle 2 is the antiparticle of 1, so that when charged, the two particles will carry the same and opposite charge, respectively. When the potential is computed in both cases and the appropriate limits are taken, we find that when the exchanged particle carries odd integer spin, like charges repel and opposite charges attract, just as in the example of electrodynamics. On the other hand, when the exchanged particle carries even integer spin, the potential is universally attractive (like charges and opposite charges attract). Hence, the spin of the graviton must be 0, 2, 4, ...

To eliminate the spin 0 possibility we note that the Eötvös experiment and recent refinements empirically indicate that gravity does couple to the energy content of objects, hence things like photons should be affected by gravity, e.g., they should "fall" in a gravitational field. If we assume that the exchanged particle is spin 0, then we lose the coupling of gravity to the spin-1 photon. Since we know that light is deflected by massive objects, e.g., the sun, then the graviton cannot be spin 0.

In qualitative field theoretic terms, the Green's functions for the propagation of the exchange particle from particle 1 to 2 in momentum space are:

$$\Delta_0 \sim \frac{1}{k^2} \qquad \text{scalar field,}$$

$$\Delta_1 \sim \frac{\eta_{\mu\nu}}{k^2} \qquad \text{vector field,} \qquad (Q.2)$$

$$\Delta_2 \sim \frac{\eta_{\mu\nu}\eta_{\sigma\rho}}{k^2} \qquad \text{tensor field.}$$

k^2 is square of the 4-momentum carried by the virtual exchange particle in the interaction, and $\eta_{\mu\nu}$ is the Minkowski flat space metric. A scalar field represents spin 0, a vector field spin 1, and an appropriately projected tensor field represents spin 2. To compute the amplitudes for the exchange, we sandwich the propagators Δ between the stress-energy tensors, $T^{\mu\nu}(1)$ and $T^{\alpha\beta}(2)$, for the two particles. For a spin-0 exchange, the propagator Δ_0 does not supply any factors of $\eta_{\mu\nu}$ in the numerator to contract the indices of $T^{\mu\nu}(1)$ with $T^{\alpha\beta}(2)$, hence we must self contract the indices separately on the two stress-energy tensors. Thus, for spin-0 exchange, the amplitude is proportional to

$$T^{\mu}{}_{\mu}(1) \, \frac{1}{k^2} \, T^{\alpha}{}_{\alpha}(2). \qquad (Q.3)$$

In other words, a spin-0 graviton only couples to the trace of the stress-energy tensor. However, the stress-energy tensor for the electromagnetic field in Minkowski space is traceless, hence scalar gravitational fields do not couple gravity to light, so the graviton cannot be spin 0.

Since the graviton is not spinless, the next possibility is spin 2. Nothing has been found classically to eliminate spin 2, so invoking the rule "if it works, don't fix it," the higher spin possibilities are neglected. Hence, the gravitational field is represented by a rank 2 tensor field. However, we are not quite finished yet. A general tensor field contains parts we still want to eliminate. For example, the antisymmetric part behaves like a couple of spin-1 fields (recall that the electromagnetic field strength $F^{\mu\nu}$ is antisymmetric) and therefore should be dumped. This leaves a symmetric tensor field.

In summary, the graviton is massless because gravity is a long ranged force and it is spin 2 in order to be able to couple to the energy content of matter with universal attraction.

In §1.2 Feynman discusses briefly the behavior of gravity and anti-matter, popularly called "antigravity." There has been only one experiment to attempt to directly measure the behavior of gravity and anti-matter for the case that the particle and antiparticle are not identical. This experiment, carried out by W. Fairbank and F. Witteborn [WiFa 67, FWML 74], was apparently motivated by DeWitt's comment at the Chapel Hill conference in 1957 (see Foreword, [DeWi 57]) where Fairbank was in attendance (see [NiGo 91] which also reviews this subject), and is an excellent example of the difficulties faced in experimental gravitational physics. In §1.2, Feynman mentions two of the arguments against antigravity based on the kaon [Good 61] and QED vacuum polarization [Schi 58,59]. If the perturbative program to quantize gravity envisioned by Feynman in these lectures yielded a consistent theory, then the issue would be settled and there would be no antigravity. Since the quantum gravity perturbation theory starts from Minkowski space, we could then expect the CPT theorem to hold at all orders and therefore the particle and antiparticle would have the same mass. In addition, the properties of the graviton discussed above would be unchanged leading to a universally attractive force including antimatter.

Unfortunately, the perturbative theory is not consistent and a lot of creative energy has been expended looking for a consistent quantized gravity. While we expect that the low energy, long distance, weak field limit of quantum gravity will be general relativity [Wein 64a,64b], Nature may require gravitational differences between matter and antimatter on very short ranges for consistency. Such effects can easily be smaller than present day experimental limits and sidestep the arguments against antigravity, but this does not mean that we will ever see anything "fall up."

Gauge Invariance and the Principle of Equivalence:

Another benefit of the field theoretic development of the theory of gravitation is that we arrive at the Principle of Equivalence, a fundamental principle at the foundation of general relativity, as a consequence of a gauge invariance. As we build a gravitational theory from the bottom up, this gauge invariance seemingly enters in an innocent way.

The free graviton is massless and moves at light speed, so we can never find a frame in which it is at rest. Therefore, there exists an invariant notion of projecting its spin onto the direction of motion and opposite the direction of motion. The massless graviton should appear with two polarizations or helicities, and no more. In general, a symmetric tensor field will have more than two *dynamic* degrees of freedom. Therefore, a rank 2 field with only two dynamical degrees of freedom is *not* a tensor, and we are in danger of losing Lorentz invariance. This situation is similar to one in electrodynamics. The way out of the dilemma is to incorporate a gauge symmetry. Hence, when we construct an action in Minkowski space to describe the free massless spin-2 gravitons, we will have to incorporate a gauge symmetry to reduce the dynamical degrees of freedom to 2. If we do not, the quantum theory will not be Lorentz invariant. The action that contains the necessary gauge symmetry and up to two derivatives of the field is the Fierz-Pauli action [FiPa 39]. This is enough to set us off and running toward general relativity (see the summary in the Foreword). We ultimately end up with the Principle of Equivalence as a result of gauge invariance. The gauge symmetry appears from the beginning in order that a quantum theory of a free massless spin-2 graviton be *Lorentz* invariant.

Battling the Infinities:

It was no secret that combining gravity and quantum mechanics was going to be a struggle. When a field is quantized, each mode of the field possesses a zero-point energy. Since the field is made up of an infinite number of modes, the vacuum energy of the quantum field is infinite. This infinity is quickly disposed of by normal ordering the field operators. The justification for doing so is that we are just redefining the zero point of the energy scale which is arbitrary to begin with. However, since gravity couples to all energy, when we add gravity, we can no longer get away with this. Vacuum fluctuations of the quantized fields do generate physical effects, so even if we cut off the number of modes, the energy density of the vacuum from the zero-point energies of the remaining modes can be quite large. Such a vacuum energy density will appear in a gravity theory as a cosmological constant. Since the cosmological constant is quite small, this is a big problem [Wein 89].

Next, the gravitational coupling constant, in units where $\hbar = c = 1$, has a dimension of (energy)$^{-2}$. Theories where the coupling constant has a positive dimension often turn out to be finite, while those whose coupling constant is dimensionless are candidates to be renormalizable. Theories with negative dimensions usually have divergences all over the place which require an infinite number of parameters to absorb the divergences and therefore are nonrenormalizable. Quantum general relativity falls into this last category.

In the process of renormalization, counterterms are generated to cancel the high energy or ultraviolet divergences that are encountered in the individual terms of the perturbation series. When the renormalization process is successful, the counterterms build a finite effective action that can be thought of as a classical field theory that contains all of the quantum effects (see, for example, [Hatf 92]). The possible counterterms are consistent with the symmetries of the original bare action. In other words, internal symmetries can severely restrict the types of counterterms that can be generated and thereby limit the number of corresponding divergences. Hence, theories with more symmetry are generally more convergent.

There are plenty of possible counterterms consistent with the known symmetries for perturbative quantum gravity, for example, terms proportional to R^2, $R^{\mu\nu} R_{\mu\nu}$, R^3, etc. Once the need for covariant ghosts was discovered, and the covariant rules for calculating perturbative terms to arbitrary order were known ([DeWi 67a, 67b], [FaPo 67]), it became obvious that Murphy's law for quantum field theory (if there is no symmetry to kill the counterterm, then the divergence will occur) was in full effect and the theory was most likely nonrenormalizable. One glimmer of hope along the way occurred when it was shown that pure quantum gravity at one-loop (the first quantum correction) is *finite* [tHVe 74], [Kore 74]. The counterterms for the Lagrangian density are

$$\mathcal{L}^{(1)} = \sqrt{g}\left[\frac{1}{120} R^2 + \frac{7}{20} R^{\mu\nu} R_{\mu\nu}\right]. \tag{Q.4}$$

At the classical level, these counterterms vanish for pure gravity since we then have $R = 0$ and $R_{\mu\nu} = 0$. However, this is not the reason that pure one-loop quantum gravity is finite. The reason it is finite on shell is that $\mathcal{L}^{(1)}$ can be absorbed at one loop by a redefinition of the metric, hence its effects are not physically observable. Recall that for pure gravity,

$$\frac{\delta \mathcal{L}^{(0)}}{\delta g_{\mu\nu}} = R^{\mu\nu} - \frac{1}{2} R g^{\mu\nu}, \tag{Q.5}$$

which, by the Principle of Least Action, generates the classical field equation for pure gravity. If we redefine the metric via

$$g'_{\mu\nu} = g_{\mu\nu} + \epsilon\, \delta g_{\mu\nu}, \qquad \delta g_{\mu\nu} \propto \frac{7}{20} R_{\mu\nu} - \frac{11}{60} R g_{\mu\nu}, \tag{Q.6}$$

then
$$\mathcal{L}^{(0)}(g) + \mathcal{L}^{(1)}(g) = \mathcal{L}^{(0)}(g') + \mathcal{O}(\epsilon^2). \qquad (Q.7)$$

$\mathcal{O}(\epsilon^2)$ terms are two-loop processes, hence one-loop is finite. When matter fields are coupled to gravity, the one-loop result is no longer finite, even at the classical level.

The hope was that there was some kind of hidden symmetry that made the one-loop result finite, and that this symmetry might render the pure gravity sector of theory finite. However, a computer calculation of two-loop corrections gave a divergent result [GoSa 86] dashing this hope. For recent reviews of the ultraviolet divergences, see [Wein 79] and [Alva 89].

One way to get improved ultraviolet behavior is to have more symmetry built into a theory. Thus, extensions or modifications to general relativity to improve the quantum behavior generally are based on additional symmetries. One popular approach is called "supergravity" (see, for example, [vanN 81]). It is based on a symmetry between bosonic and fermionic fields called "supersymmetry." When a supersymmetric theory is gauged so that the supersymmetry becomes local (a different supersymmetry transformation is allowed at each point in space-time), the gauge invariance necessarily incorporates the Principle of General Covariance and therefore gravity. In essence, each bosonic field has a supersymmetric fermionic partner and vice versa. The ultraviolet behavior of the quantum theory improves because often the ordinary divergent bosonic (fermionic) contributions from loops are cancelled by the super partner fermionic (bosonic) contributions. In other words, the supersymmetry severely restricts the type of counterterms that can be generated. Unfortunately, when the dimension of space-time is 4 there still exist potential counterterms (starting at *seven* loops in the best case). While no one knows for sure, some kind of additional or hidden symmetry or magic appears to be required to make the theory finite.

Presently, the most promising candidate theory of quantum gravity is string theory. String theory is a quantum theory in which the fundamental constituents are 1-dimensional extended objects (as opposed to point particles in ordinary quantum field theory—see, for example, [GSW 87], [Hatf 92]). If string theory is used to unify all the fundamental forces (i.e., a "theory of everything"), then the basic idea is that all matter is made up of very tiny strings whose size is of the order of the Planck length. At ordinary energy scales, such strings will be unresolved and indistinguishable from points. Unification is achieved in that all particles are made of a single kind of string, i.e., the various types of particles that we find are just different excitations of the same string. One mode of oscillation of the string is a massless, spin-2 state that can be identified as the graviton, hence string theory necessarily contains quantum gravity. The excitement over string theory results from the discovery that there exist perturbative solutions that are mathematically self-consistent or anomaly free and appear to be *finite* order by order in the perturbation series.

Intuitively the improved ultraviolet behavior of string theory arises because string theory incorporates a jumbo symmetry (modular invariance). String theory modifies point particle gravity at short distances by the exchange of massive string states which is similar to the way electroweak theory improves the ultraviolet behavior of the old 4-fermion theory of the weak interaction by replacing the 4-fermion vertex with the exchange of the massive gauge bosons $W^{\pm}$ and Z^0. The coupling constant in the old Fermi theory possesses a negative mass dimension and the theory is nonrenormalizable. The electroweak gauge theory replaces this coupling with dimensionless coupling constants associated with the boson exchange and the theory becomes renormalizable. String theory also introduces a new coupling constant, the string tension T, which in ordinary units is equivalent to an inverse square length, $L = \sqrt{c\hbar/\pi T}$. Recall that the only length that can be constructed from the gravitational constant G, $\hbar$, and the speed of light c is the Planck length, $L_p = \sqrt{G\hbar/c^3}$. The natural choice of units for the string makes the speed of light and the string tension dimensionless, $c = 1$ and $T = 1/\pi$. In these units (eliminating $\hbar$ between the expressions for L and L_p above), the gravitational constant will be dimensionless, $G = (L_p/L)^2$.

One curious feature of string theory which is quite distinct from point-particle theories is that the dimension of space-time is not an intrinsic property of the theory itself. Instead, the space-time dimension is a property of the particular solution. Anomaly-free solutions with $N = 1$ worldsheet supersymmetry can be found with a space-time dimension, D, less than or equal to the so-called critical dimension, D_c, of 10.

Unfortunately, while the individual terms in the perturbation series are finite, the sum of the series diverges [GrPe 88]. And while string theory is probably unique, the solutions to the theory certainly are not. There is no perturbative mechanism to select a particular solution or pick out the true vacuum. In this sense, the perturbative formulation of string theory loses its predictive power. Similarly, the world is not supersymmetric at ordinary energies. There is no perturbative mechanism to select solutions that admit non supersymmetric low energy spectra.

Nonperturbative Gravity:

At present it appears that there is no consistent and finite perturbative formulation of quantum gravity. In defining a perturbative expansion in general, we must make a choice for what the background metric on space-time will be from which to perturb. In a nonperturbative formulation of quantum gravity, all aspects of space-time should be determined from the solutions of the theory. For example, in string theory the propagation of the string through space-time is supposed to determine what the space-time metric is. Hence, one would prefer that the space-time metric does not appear in the formulation of the theory. For this reason and to overcome the problems mentioned above, the search for a viable nonperturbative formulation of string theory is presently underway.

With the lack of promise of perturbative quantum gravity, there is renewed interest to determine if nonperturbative quantum gravity based on general relativity makes sense. Perhaps the inconsistencies are introduced by the perturbative formulation. The canonical approach to quantum general relativity using the Wheeler-DeWitt equation and canonical variables stalled due to the complexity of the equations to be solved. Recently, however, a reformulation of general relativity in terms of new variables [Asht 86, 87] has led to a new loop representation of quantum general relativity [JaSm 88] [RoSm 90] where the equations are far easier to solve and some progress has been made. The new variables have also uncovered a relationship between general relativity and Yang-Mills which perhaps can be exploited.

Final Note: Feynman and Indices

Feynman once told me that getting minus signs, and factors of i, 2, and π down correctly was something to be bothered with only when it came time to publish the result. Apparently, index rules and conventions also fit into this category. Through the first six lectures, virtually *every* index is "down" (also see [Feyn 63b]). This produces the unconventional convention that $x_\mu = (t, z, y, x)$ and $x^\mu = (t, -z, -y, -x)$. Ignoring index rules won't work when space-time is no longer flat. I have adjusted the indices so that the standard index rules are satisfied, and have employed the nearly standard symbol $\eta_{\mu\nu}$ for the Minkowski metric (Feynman used $\delta_{\mu\nu}$).

Feynman owned a van in Pasadena that had Feynman diagrams painted on the sides and back. The van is pictured in [Syke 94]. When I first saw the van in 1981 in the parking lot at Caltech's Beckman Auditorium, I didn't know that it belonged to Feynman. It took just a few seconds to figure out it was Feynman's, because (1) the license plate had a misspelling of QUANTUM, "QANTUM," and (2) the diagram on the back, the only diagram with labels, had all indices in the down position. Feynman claimed that QED was already taken for a license plate and that QUANTUM was too long (On the other hand, I remember Feynman often spelled gauge as "guage.") On the van diagram, $\eta_{\mu\nu}$ is missing from the propagator for the photon, hence by conventional index rules the gamma matrices associated with the vertices should be labelled with one index up and the other down. After looking in one of the windows of the van and seeing a bale of hay in the back, my suspicion that the van was Feynman's was confirmed.

1.1 A FIELD APPROACH TO GRAVITATION

In this series of lectures we shall discuss gravitation in all its aspects. The fundamental law of gravitation was discovered by Newton, that gravitational forces are proportional to masses and that they follow an inverse-square law. The law was later modified by Einstein in order to make it relativistic. The changes that are needed to make the theory relativistic are very fundamental; we know that the masses of particles are not constants in relativity, so that a fundamental question is, how does the mass change in relativity affect the law of gravitation.

Now Einstein formulated his law in 1911, so that the subject is not new, and the physical results that we have to explain were beautifully explained by Einstein himself. The usual course in gravitation therefore starts by stating the laws just as Einstein did. This procedure is, however, unnecessary, and for pedagogical reasons we shall here take a different approach to the subject. Today, physics students know about quantum theory and mesons and the fundamental particles, which were unknown in Einstein's day; physics then consisted simply in gravitation and electrodynamics, and electrodynamics had forced the invention of a theory of relativity, so that the problem was to bring the theory of gravitation into line with the discoveries that had been made by studying electrodynamics.

Einstein's gravitational theory, which is said to be the greatest single achievement of theoretical physics, resulted in beautiful relations connecting gravitational phenomena with the geometry of space; this was an exciting idea. The apparent similarity of gravitational forces and electrical forces, for example, in that they both follow inverse-square laws, which every kid can understand, made every one of these "kids" dream that when he grew up, he would find the way of geometrizing electrodynamics. Thus a generation of physicists worked trying to make a so-called unified field theory, which would have unified gravitation and electrodynamics into a single thing. None of these unified field theories has been successful, and we shall not discuss them in these lectures. Most of them are mathematical games, invented by mathematically minded people who had very little knowledge of physics, and most of them are not understandable. Einstein himself worked at this, and his writings on the subject at least make some sense, but nevertheless, there is no successful unified field theory that combines gravitation and electrodynamics.

Such a success would have been short lived, however, because now we have so much more in physics than just electrodynamics and gravitation; we would need to worry about unifying mesons and kaons and neutrinos and all the other thirty or so particles that are known. Thus the unification of electrodynamics and gravitation would not have been a grand achievement, since there is so much more in the world besides electricity and gravitation.

Our pedagogical approach is more suited to meson theorists who have gotten used to the idea of fields, so that it is not hard for them to conceive that the universe is made up of the twenty-nine or thirty-one other fields all in one grand equation; the phenomena of gravitation add another such field to the pot, it is a new field which was left out of previous considerations, and it is only one of the thirty or so; explaining gravitation therefore amounts to explaining three percent of the total number of known fields.

We may even describe our approach in terms of a fiction; we imagine that in some small region of the universe, say a planet such as Venus, we have scientists who know all about the other thirty fields of the universe, who know just what we do about nucleons, mesons, etc., but who do not know about gravitation. And suddenly, an amazing new experiment is performed, which shows that two large neutral masses attract each other with a very, very tiny force. Now, what would the Venutians do with such an amazing extra experimental fact to be explained? They would probably try to interpret it in terms of the field theories which are familiar to them.

1.2   THE CHARACTERISTICS OF GRAVITATIONAL PHENOMENA

Before proceeding, let's review some of the experimental facts which a Venutian theorist would have to play with in constructing a theory for the amazing new effect.

First of all there is the fact that the attraction follows an inverse-square law. For us, this is known very, very accurately from the study of planetary orbits. Then there is the fact that the force is proportional to the masses of the objects. This was known to Galileo, who discovered all objects fall with the same acceleration. How well do we know this? Well, in principle what one should do is quite clear; we first define the mass as the inertia of an object, which we measure by applying known forces and measuring the acceleration. Then we measure the attraction due to gravitation, for example, by weighing, and compare the results. These experiments measuring forces and accelerations would be extremely difficult to perform with sufficient accuracy; but there are other ways of checking this to an accuracy of 1 part in 10^8, which was first done by Eötvös. One can do this by comparing the gravitational force of the earth with the centrifugal force due to the earth's rotation, which is a purely inertial effect. In principle, a plumb bob which is at some latitude, not $0°$ nor $90°$, does not point to the center of the earth. Actually it does not point to the center also because of the earth's bulges, but all this can be taken into account in making the comparison. In any case, at some intermediate angle such as the following: (Figure 1.1) the plumb bob points in a direction which is the resultant of the gravitational force and the centrifugal force. If we now make a plumb bob out of some other material, which has a different ratio of inertial to gravitational mass, it would hang at a slightly different angle. We can thus compare different substances; for example, if we could make the first bob out of copper, and a second one of hydrogen—come to think of it, it would be difficult to make it out of pure hydrogen—(maybe polyethylene would do), we can check the constancy of the ratio of the inertial mass to the gravitational mass

The actual experiment is not done by measuring differences in such tiny angles, but rather by measuring a torque; very small torques are more conveniently measurable because of the peculiar properties of quartz fibers, which can be made very thin and yet be able to support large weights. What is done is to hang two equal weights of the two materials on the ends of a bar, and suspend the bar in an east-west direction at its midpoint; if the component of the forces perpendicular to the gravitational forces are not equal, there is a net torque on the bar, which can be measured. The published results which I have seen, of a recent experiment by Dicke, show no effect, and the conclusion is that the ratio of the inertial masses to the gravitational masses is a constant to an accuracy of one part in 10^8, for many substances, from oxygen to lead.

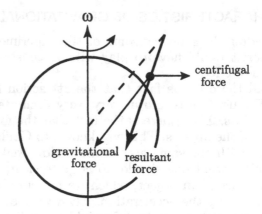

Figure 1.1

A similar experiment may be carried out by comparing the gravitational force due to the sun, to the inertial forces associated with our orbital motion about the sun. After all, here on earth we all are whirling around in space at a fantastic speed along the orbit of the earth, and the only reason we do not notice the motion is that all objects around us are also in orbits; if the gravitational attractions were not quite the same for the different objects, the objects would tend to have different orbits and there would be effects associated with this difference. The net effect would be like that of a small force in the direction of the sun. Such an effect has been looked for, in terms of a diurnal oscillation; comparing the behavior of a pair of weights on a torsion balance at dawn and at dusk, for example. Naturally, there are differences measured, some due to the fact that different sides of the building are at different temperatures— the difficulty in many of these experiments with very small effects lies in making sure that one is indeed measuring what one thinks, rather than something else. Nevertheless, the conclusion from the results is that all objects are as well balanced in their orbits as the earth is, at least to an accuracy of one part in 10^8. This accuracy of one part in 10^8 is already telling us many important things; for example, the binding energy of nuclei is typically of the order of 6 MeV per nucleon, and the nucleon mass is like 940 MeV; in short the binding energy is of the order of one percent of the total energy. Now, an accuracy of 10^8 tells us that the ratio of the inertial and gravitational mass of the *binding energy* is constant to within 1 part in 10^6. We even have a check on the ratio of the electronic binding energies for electrons in the lower shells, since 10^{-8} of a nucleon mass is something like 9 electron volts. If the experiment can be pushed to an accuracy of one in 10^{10}, as it may be in the near future, we would have a five percent check on the ratio for even chemical binding energies, which are of the order of two volts.

At these accuracies we also have a check on the gravitational behavior of antimatter. The striking similarity of electrical and gravitational forces in their inverse-square behavior, has made some people conclude that it would be nice if antimatter repelled matter; they say: in electricity, likes repel, and opposites attract; it would be nice if in gravitation it were the other way around, with likes attracting and unlikes repelling; the only candidate for gravitational "unlike" is antimatter. But with an accuracy of one in 10^8 we can check on the gravitational behavior of corrections to electronic binding energies of the K-electrons in lead due to the vacuum polarization, which involve virtual pairs and thus antimatter. It may be said at this point that there is absolutely no evidence that would require us to assume that matter and antimatter differ in their gravitational behavior. All the evidence, experimental and even a little theoretical, seems to indicate that it is the energy content which is involved in gravitation, and therefore, since matter and antimatter both represent positive energies, gravitation makes no distinction.

Further evidence comes from the fact that light "falls" in a gravitational field by an amount which is given by our theory; it is deflected as it goes by the sun, for example, by a measurable amount, which we shall calculate later. But the photon is its own antiparticle, so that we must conclude that both particle and antiparticle in this case show identical gravitational behavior. It might be amusing as an exercise to some people to construct a theory in which photons that came from an electron were different from photons that came from a positron, but since there is absolutely no evidence that such a thing is necessary to explain any phenomena, there is very little point in trying to construct such a theory; it would have to explain all known things as well as the present theory—and it is most likely that it could be proved wrong by some other effect, in that it would possibly predict some new electrodynamic effects which experiment would quickly show absent.

The most direct evidence that matter and antimatter do behave identically with regard to gravitation comes from the experiments on the decay of the K_0 and $\bar{K}_0$ done at M.I.T. The experiment itself is not without its flaws, but perhaps we may use the results to kill the theory of unequal behavior. These arguments are due to M. Good [Good 61].

We suppose that the K_0 and $\bar{K}_0$ are affected by gravity, otherwise the argument does not work. These two are antiparticles of each other, so we will see what happens if one is attracted, and the other is repelled by gravity. These particles have two decay modes which are describable as

$$K_1 = \frac{1}{\sqrt{2}}\left(K_0 + \bar{K}_0\right), \qquad K_2 = \frac{1}{\sqrt{2}}\left(K_0 - \bar{K}_0\right).$$

The amplitudes for decay by these two modes interfere; the experiment has detected this interference, and assigned a value Δm to the mass difference, such that $\Delta m < \hbar/10^{-10}$ sec. This value is inconsistent with the idea that matter is attracted but antimatter repelled, because of the fact that the experiment was carried out in the gravitational field of the earth; if the gravitational potential is ϕ, then there is an increase or decrease in mass by $m\phi$ for one, and $-m\phi$ for the other; the expected mass difference is larger than the limit set by the M.I.T. experiment. If we consider not the earth's gravitational potential, but the sun's, which is larger, or even the Galaxy's, we get better and better limits on the degree to which the equality of gravitational interaction holds. The whole argument can be brushed aside by those who cling to the antimatter-repels theory; all that is needed is that K_0 and $\bar{K}_0$ should not be gravitating particles, but this requires a new special assumption. It is evident that any single experimental fact can be disregarded if we are willing to think up a special reason why the experiment should show the result that it does.

It is also known that single free neutrons fall in a gravitational field as might be expected. This is known to fair accuracy because it must be taken into account in designing neutron interferometers; slow neutrons from a pile may be collimated into narrow beams, and detected some distance away, of the order of hundreds of feet; it is found that they fall in the earth's gravitational field like any other particles we can measure. In summary, the first amazing fact about gravitation is that the ratio of inertial mass to gravitational mass is constant wherever we have checked it.

The second amazing thing about gravitation is how weak it is. It is in fact so weak that if the Venutians had called the β-decay interactions the "Weak" interactions, the discovery of gravitation would be a tremendous embarrassment. Evidently, gravitation plays an important role in our lives, so that gravitational forces are not weak compared to the strength of our leg muscles; we mean that gravitational forces are very weak compared to the other forces that exist between particles. This comparison is presumably something more "universal" than a comparison with something of human strength. Let us, for example, compute the ratio of gravitational to electrical force between two electrons. The result is

$$F_{\text{electricity}}/F_{\text{gravity}} = 4.17 \times 10^{42}.$$

In other words, gravitation is *really* weak. This comparison in terms of the ratio of forces is more significant than the usual statement in terms of coupling constants; for example, it is often said that electromagnetic forces are "weak" because the number $e^2/\hbar c$ is small, namely $1/137$. But quoting $1/137$ is not very significant, since we could just as well imagine that the most significant number to quote is a rationalized dimensionless charge for the electron, which would be $\sqrt{(4\pi e^2/\hbar c)} = 0.31$, which looks very different from $1/137$ but has the same physical content. Thus, when the weak (β-decay) interaction is said to be weak because its coupling strength is a "small" $GM_p^2 = 10^{-5}$, we may ask, why is the proton mass involved here? If the weak interactions are mediated by some meson, called the B meson these days, it might be more natural to involve the square of the B meson mass, which might be much larger than the nucleon mass, enough to give a coupling constant very different from the "small" 10^{-5}.

All other fields that we know about are so much stronger than gravity that we have a feeling that gravitation would probably never be explained as some correction, some left-over terms until now neglected, in a theory that might unify all other fields that we know about. The number 10^{42} is so spectacularly large that it is very tempting to go looking for other large numbers that might be associated with it, an idea originally due to Eddington [Eddi 31,36,46]. One such large number is puzzling enough, and two large numbers would be worse; it might make us feel better to know that they were connected, so that the largeness of one implies the largeness of the other; thus leaving one large thing to be explained rather than two. Eddington proposed that $10^{42} = 2^{137}$, but some parts of his book are so obscure that it may be said not to contain a useful theory at all; it is simply too obscure. We shall look for an explanation of 10^{42} in another direction. We know of other such large numbers, for example, the number of atoms or of particles in us, but again, we would like to get away from our human nature in making these comparisons. It is interesting that gravitational forces are involved in the motion of the largest objects, such as galaxies, so we might look for a connection between the strength of gravitational forces and the size of the universe.

Now the universe is very large, and its boundaries are not known very well, but it is still possible to define some kind of a radius to be associated with it. It is a matter of observation that the light from distant stars and galaxies is shifted toward lower frequencies as though they were receding with a velocity proportional to their distance from us. This may be explained by the so-called explosion theory of the universe. As we shall see, the theory of gravitation is very important in considering cosmological models, and we shall discuss them later in our course. But for now suppose that galaxies are made from stuff that got started from a given spot in a big explosion; then the proportionality between velocity from the center and distance follows very naturally, since the stuff which is farther out is farther out because it is going faster. The proportionality is such that $R/V = T$, the time since the hypothetical start. This time, (its inverse is known as Hubble's constant) has a value something like 13×10^{10} years. The error in this value is a sizeable fraction; some years ago the value was quoted as 2×10^{10} years. The errors come from the measurements of the distances; the Doppler shifts are much more easily measured than the distances to faraway galaxies.

This constant represents a lifetime of the universe; not necessarily that we believe that the universe did begin T years ago, but rather it represents a fundamental dimension of the universe, much in the way that the quantity e^2/mc^2 represents the "electron radius." In the same way, the quantity Tc represents a length which may be called the "radius" of the universe. Let us see whether we can get a factor of the order of 10^{42} somewhere from Hubble's constant. We may for example take the ratio of the time that light takes to go the Compton wavelength of an electron $\hbar/mc^2$, or of a proton, $\hbar/M_pc^2$, to the Hubble constant; again, we hope that this ratio represents something more significant than a number of seconds, which is a human scale. These times, $\hbar/mc^2 = 10^{-21}$ sec., or $\hbar/M_pc^2 = 10^{-24}$ sec., and the time T is $T = 10^{17}$ sec.; the ratio is like 10^{41} for protons, which for these numerological arguments is not too far from the ratio of electric to gravitational force for protons, near 10^{36}. If we use electrons, the ratio is 10^{38}, not very close to 10^{42}, but we must keep in perspective what it is we are doing, simply to put forth some wild speculations in order to see whether we get any real ideas. As a matter of fact, Dirac [Dira 37,38] has tried to work out some theory of gravitation in which the gravitational constant is precisely this number. One of the difficulties is that he must assign a time dependence to the strength of gravitation, since the universe keeps getting older in units of the Compton wave time. But it is very difficult to define what one means by saying that the forces of gravitation are time-dependent while every thing else "remains the same." Since the significant numbers are the dimensionless ratios of things, he might just as well describe the situation by saying that the electric charge is time dependent, so that his theory is really not well defined. Let us for the moment overlook these difficulties and see whether intuitively we can deduce any consequences of the time dependence of the gravitational constant. Some people say that this explains earthquakes; the earth is stretching slowly as the gravitational forces weaken, so that we expect cracks and what-not. But the alternative theory, that there are currents in the magmatic interior of the earth, explains better why the earthquakes occur in highly localized regions of the earth's crust. So that there is little comfort to be found in the theory of earthquakes.

One might try to disprove this idea that G changes from consequences in the theory of stars; we shall not study stars in detail, but the idea of how they work is that, as stellar material falls into a center, it acquires gravitational energy and heats up to the point that nuclear reactions can take place, and the pressure then keeps the star in an equilibrium situation, the energy lost by radiation now being supplied by nuclear reactions, which does not allow it to collapse further. If we now assume that the gravitational constant is time dependent, being larger in the past, we must assume that the rate of energy generation was higher in the past to support the greater weight; a detailed consideration shows that we might expect the luminosity of a star to go roughly as the sixth power of the magnitude of the gravitational constant; qualitatively, if the constant is larger, higher central temperature is needed to support the greater weight, so that the nuclear reactions proceed at a faster rate. We may ask what effect this would have on our sun and thence on the surface temperature of the earth; we claim to have some evidence on the surface temperature of the earth from the fact that we know there has been some kind of life on our planet for some 10^9 years. If the gravitational constant had been larger then, the luminosity of the sun would be greater according to a G^6 law, and the earth would have been in orbit closer to the sun than it is now. The light intensity incident on the earth would have been proportional to G^8. Now we can make some estimate of the earth's temperature; a real calculation is difficult, because the earth is shiny and has clouds and all sorts of complications, but we can estimate by assuming it is a blackbody. The blackbody radiates energy as T^4, from its entire surface, since it is rotating, and it comes to some equilibrium temperature with the incident energy from the sun; if we compare the answer to the measured temperatures for the surfaces of the planets, the answer turns out to be surprisingly good (it fits everywhere that the surface temperature is known), so we can use this estimate in figuring the temperature of the earth's surface some billion years ago, with a gravitational constant of the order of 8 percent larger. The incident energy goes as T^4 and E goes as G^8, so that the earth's temperature is proportional to G^2, or some 16 percent, or 48°C, higher a billion years ago.

We may now ask the geophysicists and biochemists what the surface of the earth would be like at some temperature such as 75°C. It is not quite hot enough to make the seas boil, so we can't get rid of our theory entirely. It is conceivable that life did get started at such temperatures of the water. We have here on earth places, such as the hot springs in Yellowstone, where certain bacteria live in water at similar temperatures. It would be a strange life that could exist, and the oldest fossils found on earth do not suggest any peculiarities that would be reasonable evidence for such higher temperatures; nevertheless, as far as I know, we can't claim that there is conclusive evidence against a higher earth temperature in early times.

The very much greater luminosity of stars if the gravity constant were greater in the past would change the evolutionary time scale of certain stars. I know that some astronomers are trying to see if this is inconsistent with observations, but I don't know if they have arrived at a sharp conclusion yet.

Another spectacular coincidence relating the gravitational constant to the size of the universe comes in considering the total energy. The total gravitational energy of all the particles of the universe is something like GMM/R, where $R = Tc$, and T is Hubble's time. Actually, if the universe were a sphere of constant density, there would be a factor of 3/5, but let us neglect this, since our cosmological model is not all that well known.

If now we compare this number to the total rest energy of the universe, Mc^2, lo and behold, we get the amazing result that $GM^2/R = Mc^2$, so that the total energy of the universe is *zero*. Actually, we don't know the density nor that radius well enough to claim equality, but the fact that these two numbers should be of the same magnitude is a truly amazing coincidence. It is exciting to think that it costs *nothing* to create a new particle, since we can create it at the center of the universe where it will have a negative gravitational energy equal to Mc^2.

In these estimates, it is the density of the universe that is the hardest to determine. We can see stars and galaxies, sure enough, but we have no clear idea of how many black stars there are, stars which have burned out. Neither do we know the density of intergalactic gas. We have some idea of the density of sodium in the space between galaxies, obtained by measuring the absorption of the D lines as they come from distant stars. But sodium is possibly a very tiny fraction of the total, and we would need to know the hydrogen density. By studying the motion of spiral arms of galaxies, and of globular clusters, it appears that galaxies have in their center a lump of dark mass. All of this fragmentary evidence does not allow us to get any reliable estimate of the average density of the universe. Eddington, for his estimates in the twenties, used the value 1 hydrogen atom per cm^3, for galaxies. The radio astronomers who have recently studied the galaxy in "hydrogen light" arrive at a value only slightly smaller, say 0.7 hydrogen atoms/cm^3. There is no evidence for the density of intergallactic material; the cosmologists have guessed at values some 10^5 times smaller, 10 hydrogen atoms per cubic meter. With this estimate, we get the exciting result that the total energy of the universe is zero. Why this should be so is one of the great mysteries— and therefore one of the important questions of physics. After all, what would be the use of studying physics if the mysteries were not the most important things to investigate?

All of these speculations on possible connections between the size of the universe, the number of particles, and gravitation, are not original but have been made in the past by many other people. These speculators are generally of one of two types, either very serious mathematical players who construct mathematical cosmological models, or rather joking types who point out amusing numerical curiosities with a wistful hope that it might all make sense some day.

1.3   QUANTUM EFFECTS IN GRAVITATION

In the next few lectures we shall start to construct a quantum theory of gravitation. It might be well for us to keep in mind whether there would be any observable effects of such a theory. Let us first consider the gravitation as a perturbation on the hydrogen atoms. Evidently, an extra attraction between the electron and proton produces a small change in the energy of bound hydrogen; we can calculate this energy change from perturbation theory and get a value, ϵ. Now, the time dependent wave function of a hydrogen atom goes as $\psi = \exp(-iEt)$, with E of such a size that the frequency is something like 10^{16} cycles per second. Now, in order to observe any effects due to ϵ, we should have to wait for a time until the true wave function should differ from the unperturbed wave function by something like 2π in phase. But the magnitude of ϵ is so small that the phase difference would be only 43 seconds (of phase) in a time equal to 100 times the age of the universe T. So gravitational effects in atoms are unobservable.

Let us consider another possibility, an atom held together by gravity alone. For example, we might have two neutrons in a bound state. When we calculate the Bohr radius of such an atom, we find that it would be 10^8 light years, and that the atomic binding energy would be 10^{-70} Rydbergs. There is then little hope of ever observing gravitational effects on systems which are simple enough to be calculable in quantum mechanics.

Another prediction of the quantum theory of gravitation would be that the force would be mediated by the virtual exchange of some particle, which is usually called the graviton. We might therefore expect that under certain circumstances we might see some gravitons, as we have been able to observe photons. But let us recall that even though light has been observed very early in man's history (Adam did it) it was not until 1898 that electromagnetic waves were produced with conscious knowledge of their field nature, and that the quantum aspects of these waves were not observed until even later. We observe gravity, in that we know we are pulled to the earth, but classical gravitational waves have not as yet been observed; this is not inconsistent with what we expect—gravitation is so weak that no experiment that we could perform today would be anywhere near sensitive enough to measure gravitational radiation waves, at least, those which are expected to exist from the strongest sources that we might consider, such as rapidly rotating double stars. And the quantum aspect of gravitational waves is a million times further removed from detectability; there is apparently no hope of ever observing a graviton.

1.4   ON THE PHILOSOPHICAL PROBLEMS IN QUANTIZING MACROSCOPIC OBJECTS

The extreme weakness of quantum gravitational effects now poses some philosophical problems; maybe nature is trying to tell us something new here, maybe we should not try to quantize gravity. Is it possible perhaps that we should not insist on a uniformity of nature that would make everything quantized? Is it possible that gravity is not quantized and all the rest of the world is? There are some arguments that have been made in

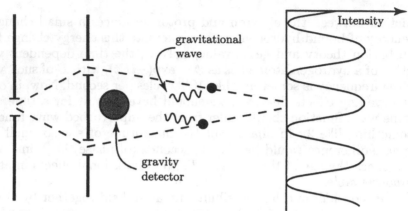

Figure 1.2

the past that the world cannot be one-half quantum and one-half classical. Now the postulate that defines quantum mechanical behavior is that there is an amplitude for different processes. It cannot then be that a particle which is described by an amplitude, such as an electron, has an interaction which is not described by an amplitude but by a probability. We consider a two-slit diffraction experiment, and insert a gravity detector, which may be assumed classical, which can in principle tell which slit the electron passed as illustrated in Figure 1.2. Let us imagine that the detector has not yet received a signal telling it which slit the electron passed; the position of the electron is described by an amplitude, half of which went through the upper slit, and half of which went through the lower slit. If the gravity interacts through a field, it follows that the gravity field must have an amplitude also; half of which corresponds to the gravity field of an electron which went through either slit. But this is precisely the characteristic of a quantum field, that it should be described by an amplitude rather than a probability! Thus it seems that it should be impossible to destroy the quantum nature of fields.

In spite of these arguments, we would like to keep an open mind. It is still possible that quantum theory does not absolutely guarantee that gravity *has* to be quantized. I don't want to be misunderstood here—by an open mind I do not mean an empty mind—I mean that perhaps if we consider alternative theories which do not seem *a priori* justified, and we calculate what things would be like if such a theory were true, we might all of a sudden discover that's the way it really is! We would never make this discovery with the attitude that "of course one must always entertain the possibility of doubt," but act and calculate with one prejudice only. In this spirit, I would like to suggest that it is possible that quantum mechanics fails at large distances and for large objects. Now, mind you, I do not say that I think that quantum mechanics *does* fail at large distances, I only say that it is not inconsistent with what we do know. If this failure of quantum mechanics is connected with gravity, we might speculatively expect this to happen for masses such that $GM^2/\hbar c = 1$, of M near 10^{-5} grams, which corresponds to some 10^{18} particles. Now quantum mechanics gives silly answers for objects of this size; if we calculate the probability that a grain of sand should jump over a wall, we get answers like $10^{-260,000}$, which are ridiculous. We must therefore not neglect to consider that it is possible for quantum mechanics to be wrong on a large scale, to fail for objects of ordinary size. In this connection we might discuss how the theory of observation and measurement creates some problems. Let us for example talk about Schrödinger's cat paradox. It is not a real paradox, in the sense that there are two possible answers which can be arrived at by proper logic—it is a means of pointing out a philosophical difficulty in quantum mechanics, and each physicist must decide which side he prefers.

We imagine that in a closed box into which we cannot observe, we have placed a live cat and a loaded shotgun; the cat is confined so that if the shotgun goes off it dies. Now the shotgun is triggered by a Geiger counter, which counts particles from a radioactive source; let us suppose that the source is such that we expect one count per hour. The question is, what is the probability that the cat is alive one hour after we have closed the box?

The answer from quantum mechanics is quite easy; there are two possible final states that we consider, the amplitude is

$$\text{Amplitude} = \frac{1}{\sqrt{2}}\,\psi(\text{cat alive}) + \frac{1}{\sqrt{2}}\,\psi(\text{cat dead}).$$

When we think about this answer, we have the feeling that the cat does not see things in the same way; he does not feel that *he* is $1/\sqrt{2}$ alive and $1/\sqrt{2}$ dead, but one or the other. So that what may properly be described by an amplitude to an external observer, is not necessarily well described by a similar amplitude when the observer is part of the amplitude. Thus the external observer of the usual quantum mechanics is in a peculiar position. In order to find out whether the cat is alive or dead, he makes a little hole in the box and looks; it is only after he has made his measurement that the system is in a well-defined final state; but clearly, from the point of view of the internal observer, the results of this measurement by the external observer are determined by a probability, not an amplitude. Thus we see that in the traditional description of quantum mechanics we have a built-in difference between a description including an external observer, and a description without observation.

This kind of a paradox crops up each time that we consider the amplification of an atomic event so that we recognize how it affects the whole universe. The traditional description of the total quantum mechanics of the world by a complete Monster Wavefunction (which includes all observers) obeying a Schrödinger equation

$$i \frac{\partial \Psi}{\partial t} = \mathbf{H}\Psi$$

implies an incredibly complex infinity of amplitudes. If I am gambling in Las Vegas, and am about to put some money into number twenty-two at roulette, and the girl next to me spills her drink because she sees someone she knows, so that I stop before betting, and twenty-two comes up, I can see that the whole course of the universe for me has hung on the fact that some little photon hit the nerve ends of her retina. Thus the whole universe bifurcates at each atomic event. Now some people who insist on taking all quantum mechanics to the letter are satisfied with such a picture; since there is no outside observer for a wavefunction describing the whole universe, they maintain that the proper description of the world includes all the amplitudes that thus bifurcate from each atomic event. But nevertheless, we who are part of such a universe know which way the world has bifurcated for us, so that we can follow the track of our past. Now, the philosophical question before us is, when we make an observation of our track in the past, does the result of our observation become real in the same sense that the final state would be defined if an outside observer were to make the observation? This is all very confusing, especially when we consider that even though we may consistently consider ourselves always to be the outside observer when we look at the rest of the world, the rest of the world is at the same time observing us, and that often we agree on what we see in each other. Does this then mean that my observations become real only when I observe an observer observing something as it happens? This is a horrible viewpoint. Do you seriously entertain the thought that without the observer there is no reality? Which observer? Any observer? Is a fly an observer? Is a star an observer? Was there no reality in the universe before 10^9 B.C. when life began? Or are *you* the observer? Then there is no reality to the world after you are dead? I know a number of otherwise respectable physicists who have bought life insurance. By what philosophy will the universe without man be understood?

In order to make some sense here, we must keep an open mind about the possibility that for sufficiently complex processes, amplitudes become probabilities. The fact that it is amplitudes that are being added could be detectable only by processes which detect phase differences, and interferences. Now the phase relations for very complicated objects could be enormously complex, so that one would observe an interference only if the phases of all parts of a complicated object were to evolve in a very, very precise fashion. If there were some mechanism by which the phase evolution had a little bit of smearing in it, so it was not absolutely precise, then our amplitudes would become probabilities for very complex objects. But surely, if the phases did have this built-in smearing, there might be some consequences to be associated with this smearing. If one such consequence were to be the existence of gravitation itself, then there would be no quantum theory of gravitation, which would be a terrifying idea for the rest of these lectures.

These are very wild speculations, and it would be little profit to keep discussing them; we should always keep in mind the possibility that quantum mechanics may fail, since it has certain difficulties with the philosophical prejudices that we have about measurement and observation.

1.5   GRAVITATION AS A CONSEQUENCE OF OTHER FIELDS

Let us return to a construction of a theory of gravitation, as our friends the Venutians might go about it. In general we expect that there would be two schools of thought about what to do with the new phenomenon. These are:

1.   That gravitation is a new field, number 31.
2.   That gravitation is a consequence of something that we already know, but that we have not calculated correctly.

We shall take up the second point of view for a little while, to see whether it has any possibilities. The fact of a universal attraction might remind us of the situation in molecular physics; we know that all molecules attract one another by a force which at long distances goes like $1/r^6$. This we understand in terms of dipole moments which are induced by fluctuations in the charge distributions of molecules. That this is universal is well known from the fact that *all* substances may be made to condense by cooling them sufficiently. Well, one possibility is that gravitation may be some attraction due to similar fluctuations in something, we do not know just what, perhaps having to do with charge.

If we worry about the fact that quantum mechanics fails in that very often infinities crop up in summing over all states, we might look for a connection between gravity, the size of the universe, and this failure of quantum mechanics. The infinities always occur when we sum over denominators $\sum_n 1/(E - E_n)$. Now, it is conceivable that if we were to consider that whole universe, we would not be summing over virtual states in the usual fashion, but that we should sum only over those virtual states for which we could borrow enough energy from the rest of the universe. A theory that would not allow virtual states if the energy violation were larger than the total energy of the universe would be slightly different from the usual one, which assumes this total energy is essentially infinite.

There would be differences from the usual theory, but I suspect that nothing like gravitation will result as a consequence from such a theory.

We might consider whether gravitational forces might not come from the virtual exchange of a particle which is already known, such as the neutrino. After all, superficially it has the right qualities, since it is a neutral particle with zero mass, so that its interaction would go like $1/r$, and its interaction will be very weak.

In the next lecture, we shall pursue this neutrino theory of gravitation, and find out how it fails. Then we shall begin to construct a theory of gravitation as the 31st field to be discovered.

2.1 POSTULATES OF STATISTICAL MECHANICS

In constructing our theories of gravitation, we should be wary about accepting too glibly many of the prejudices of the present scientific thinking. In the last lecture we saw how there was something unsatisfactory in the way that probabilities appear in our interpretation of the universe. If we truly think that the universe is described by a grand wavefunction without an outside observer, nothing can ever become a probability, since no measurement is ever made! This many physicists hold in spite of our experimental evidence in dealing with subregions of the universe, which are for our purposes very accurately described by wavefunctions which represent amplitudes for the probabilities of results of measurements.

In a similar fashion, there are difficulties with the usual simple text book description of statistical mechanics; although not closely related to the theory of gravitation it is related to cosmological questions which we shall discuss later. Often one postulates that *a priori*, all states are equally probable. This is not true in the world as we see it. This world is not correctly described by the physics which assumes this postulate. There are in the world people—not physicists—such as geologists, astronomers, historians, biologists, who are willing to state high odds that when we look into an as yet unobserved region of the universe, we shall find certain

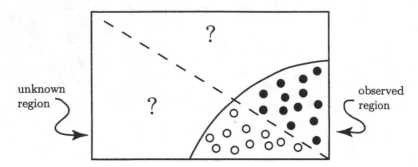

Figure 2.1

organization which is not predictable by the physics we profess to believe. In our experience as observers, we find that if we look inside a book which says "Napoleon" on the cover, the odds are indeed very great that there will be something about Napoleon on the inside. We certainly do not expect to find a system in thermodynamic equilibrium as we open the pages. But the physicists have not found a way of incorporating such odds into unobserved regions of the universe. The present physics would never predict as well as a geologist the odds that when we look inside certain rocks, we will find a fossil.

In the same way, astronomical historians and astronomers find that everywhere in the universe that we look we see stars, which are hotter on the inside and cooler on the outside, systems which are indeed very far from thermodynamic equilibrium.

We can get an idea of how unlikely this situation is, from the point of view of the usual prejudices of thermodynamics, by considering simple orderings. Consider a box as the universe, in which there are two kinds of particles, white and black. Suppose that in a certain region of the universe, such as a small corner of a box, we see that all white particles are separated from the black particles along the diagonal (Figure 2.1). The *a priori* probability that this should be is very, very small, since we must, with our present prejudices, ascribe it to a statistical fluctuation which is rather unlikely. What do we predict for the rest of the universe? The prediction is, that as we look into the next region, we should most likely find that the next region has a much less ordered distribution of the white and black particles. But in fact we do not find this, but each new region that we see, we see the same order as before. If I get into my car and drive to the mountains I have never seen previously, I find trees that look just like the trees that I know. With the extremely large numbers of particles that are involved, the probabilities of such observations in terms of the usual statistical mechanics are fantastically small.

The best explanation of a fluctuation that one observes is that only that much has fluctuated and that the rest of the stuff is at random. If all states were equally likely *a priori*, and if one found a piece of the world lopsided, then the rest of the world should be uniformly mixed because it would be less of a fluctuation. It may be objected that the events and structures in the world are correlated; they all have the same past! But that is a different theory than the one underlying the statistical mechanical description of the universe. There is this opposite theory, which is that in the past the world was more organized than now, and that the most likely state is not that of equilibrium, but some special state which is dynamically evolving. This is the common sense assumption which all historians, paleontologists, and others adopt.

Probability arguments must be used as a test for the theory, and may be employed as follows. Let us suppose that on *a priori* grounds we wish to assign very, very low odds to the hypothesis that the universe is not to be described as an elaborate fluctuation from the complete chaos characterizing thermodynamic equilibrium; for example, let us suppose that the *a priori* probability for the idea that all states are equally likely is $1 - 10^{-100}$. Then let us describe a number of ordered states according to some scheme; for example, suppose that we list all states whose ordering can be specified by less than a million words. Now we assign the remaining *a priori* probability of 10^{-100} to the hypothesis that the universe is evolving from one of these special ordered conditions in the past. In other words, we adopt the prejudice that all states are equally likely, but wish to allow the possibility that observational tests might invalidate the equilibrium hypothesis.

Now we begin to make observations on the world about us, and we observe states with a describable ordering. Each one of us this morning saw that the ground was below and the air was above, but one such observation is enough to increase greatly the odds for the ordered states in the *a posteriori* judgment on the probability of the initial situation. And as we make more and more observations, this increase eventually overtakes even the 10^{-100}, in a way which may be computed according to a theorem: if the *a priori* probability of situation A is P_a and the *a priori* probability of situation B is P_b, and if an observation is made which is more likely if A holds and less likely if B holds, the *a posteriori* probability of A increases by the ratio by which the result of the measurement is more probable if A holds.

If one makes an observation of a corner of the universe, anything macroscopic, one finds that it is far removed from equilibrium. The odds that this should be a fluctuation are extremely small; it requires only a single observation of macroscopic order to reduce the probability to 10^{-2000}, for which only 5000 molecules have to be ordered. Thus, it is perfectly obvious that only special states could possibly give rise to the immense degree of order we see in the world.

How then does thermodynamics work, if its postulates are misleading? The trick is that we have always arranged things so that we do not do experiments on things as we find them, but only after we have thrown out precisely all those situations which would lead to undesirable orderings. If we are to make measurements on gases, which are initially put into a metal can, we are careful to "wait until thermodynamic equilibrium has set in," (how often have we heard *that* phrase!) and we throw away all those situations in which something happens to the apparatus, that the electricity goes off because a fuse is blown, or that someone hits the can with a hammer. We never do experiments on the universe as we find it, but rather we control things to prepare rather carefully the systems on which we do the experimenting.

A more satisfactory way of presenting the postulates of statistical mechanics might go as follows. Suppose that we do know all details of a (classical) system, such as a mass of gas with infinite precision; that is, we know the positions and velocities of all the particles at some instant of time $t = 0$. We can then (disregarding the actual difficulties in practice) calculate exactly if we know the laws of nature exactly, and find out the behavior and state of all other particles at any time in the future. But now suppose that there is some small uncertainty in our measurements, or in our knowledge of any one fact that enters into the calculation, the position, the velocity, of any one particle, or a small uncertainty in the exactness with which we know the interactions of the particles. It does not at all matter (excepting counterexamples designed by mathematicians) where the uncertainty is. If such an uncertainty exists, we shall have to describe the final state by averaging over the uncertainty, and if a sufficiently long time elapses, which will be shorter the larger the uncertainty and the larger the system, the predictions of measurements will be very close to those given by the canonical theory of thermodynamic equilibrium.

If for example we plot the velocity of molecule number 6 at the time $t = 30$ min. as a function of any other starting variable in the system, such as the initial position or velocity of particle number 133, we shall find an extremely complicated curve of very, very fine detail, which should average to the "equilibrium" results as soon as we average over the initial finite uncertainty of the variable in question. In other words, the distribution of final values in the range considered should be much like the "equilibrium" distribution (Figure 2.2).

A physically satisfactory discussion of thermodynamics and statistical mechanics can only be achieved if it is recognized that the problem is to define conditions in a system in which various things happen at very different rates. Only if the rates are sufficiently distinct is thermodynamics of use. Thus the theory of thermodynamics must distinguish between slow and fast processes. When we are talking about thermodynamic equilibrium for our mass of gas, we do not wait for an infinite time, but for a time very long compared to a certain class of interaction (for example molecular collisions) which is producing the type of equilibrium we are considering. In studying oxygen in a metal can, we do not wait so long

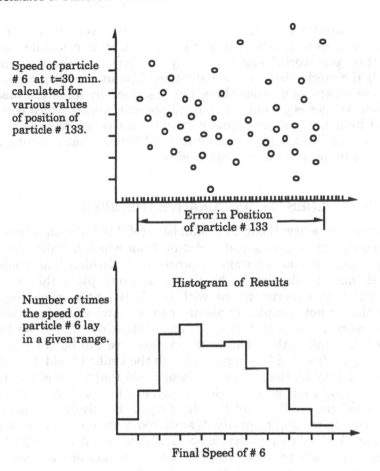

Speed of particle # 6 at t=30 min. calculated for various values of position of particle # 133.

Error in Position of particle # 133

Histogram of Results

Number of times the speed of particle # 6 lay in a given range.

Final Speed of # 6

Figure 2.2

that the walls of the can have oxidized, or that the metal has evaporated into space, as it eventually should do, since it has a finite vapor pressure, nor do we consider all the nuclear reactions that once in a very great while do (according to our theory) occur for the colliding molecules.

We must be careful to interpret the results of our theories when they are treated with full mathematical rigor. We do not have the physical rigor sufficiently well defined. If there is something very slightly wrong in our definition of the theories, then the full mathematical rigor may convert these errors into ridiculous conclusions.

The question is how, in quantum mechanics, to describe the idea that the state of the universe in the past was something special. The obvious way is to say that the wave function of the world (if such exists) was a certain ψ_0 at $t = -$ (age of the universe). But that means the wave function ψ at present tells us not only about our world but equally of all the other possible universes that could have evolved from this same beginning. This is the cat paradox on a large scale. Equally represented is "our world" plus all other dead cats whose death was a quantum controlled accident. From this "our world" can be got by "reduction of the wave packet." What is the mechanism of this reduction? You must either suppose that observing creatures do something not described by quantum mechanics (i.e., Schrödinger equation) or that all possible worlds which could have evolved from the past are equally "real." This is not to say that either choice is "bad," but only to point out that I believe that present quantum mechanics implies one or the other idea.

2.2 DIFFICULTIES OF SPECULATIVE THEORIES

In constructing a new theory, we shall be careful to insist that they should be precise theories, giving a description from which definite conclusions can be drawn. We do not want to proceed in a fashion that would allow us to change the details of the theory at every place that we find it in conflict with experiment, or with our initial postulates. Any vague theory that is not completely absurd can be patched up by more vague talk at every point that brings up inconsistencies—and if we begin to believe in the talk rather than in the evidence we will be in a sorry state. Something of this kind has happened with the Unified Field Theories. For example, it may be that one such theory said that there is a tensor $J_{\mu\nu}$ which "is associated" with the electromagnetic tensor. But what does this "associated" mean? If we set the thing equal, the theory predicts wrong effects. But if we don't specify "associated," we don't know what has been said. And talk that this "association" is meant to "suggest" some new relation leads to nowhere. The wrong predictions are ascribed to the wrong "suggestions" each time, rather than to the wrong theory, and people keep thinking of adding a new piece of some antisymmetric tensor which would somehow fix things up. This speculative talk is no more to be believed than the talk of numerologists who find accidental relations between certain magnitudes, which must be continuously modified as these magnitudes are measured with more precision, first relating units, and then smaller and smaller fractions of these units to keep up with the smaller and smaller uncertainties in the measured values.

In this connection I would like to relate an anecdote, something from a conversation after a cocktail party in Paris some years ago. There was a time at which all the ladies mysteriously disappeared, and I was left facing a famous professor, solemnly seated in an armchair, surrounded by his students. He asked, "Tell me, Professor Feynman, how sure are you that the photon has no rest mass?" I answered "Well, it depends on the mass; evidently if the mass is infinitesimally small, so that it would have no effect whatsoever, I could not disprove its existence, but I would be glad to discuss the possibility that the mass is not of a certain definite size. The condition is that after I give you arguments against such mass, it should be against the rules to change the mass." The professor then chose a mass of 10^{-6} of an electron mass.

My answer was that, if we agreed that the mass of the photon was related to the frequency as $\omega = \sqrt{k^2 + m^2}$, photons of different wavelengths would travel with different velocities. Then in observing an eclipsing double star, which was sufficiently far away, we would observe the eclipse in blue light and red light at different times. Since nothing like this is observed, we can put an upper limit on the mass, which, if you do the numbers, turns out to be of the order of 10^{-9} electron masses. The answer was translated to the professor. Then he wanted to know what I would have said if he had said 10^{-12} electron masses. The translating student was embarrassed by the question, and I protested that this was against the rules, but I agreed to try again.

If the photons have a small mass, equal for all photons, larger fractional differences from the massless behavior are expected as the wavelength gets longer. So that from the sharpness of the known reflection of pulses in radar, we can put an upper limit to the photon mass which is somewhat better than from an eclipsing double star argument. It turns out that the mass had to be smaller than 10^{-15} electron masses.

After this, the professor wanted to change the mass again, and make it 10^{-18} electron masses. The students all became rather uneasy at this question, and I protested that, if he kept breaking the rules, and making the mass smaller and smaller, evidently I would be unable to make an argument at some point. Nevertheless, I tried again. I asked him whether he agreed that if the photon had a small mass, then from field theory arguments the potential should go as $\exp(-mr)/r$. He agreed. Then, the earth has a static magnetic field, which is known to extend out into space for some distance, from the behavior of the cosmic rays, a distance at least of the order of a few earth radii. But this means that the photon mass must be of a size smaller than that corresponding to a decay length of the order of 8000 miles, or some 10^{-20} electron masses. At this point, the conversation ended, to my great relief.

We do not want to do similar things in attempting to construct a theory of gravitation from the *known* fields, modifying the sizes of couplings or introducing new postulates at every point that we find a difficulty; we should be prepared to put forth definite theories using the *known* behavior of our fields, and be prepared to reject them if they are inadequate.

2.3 THE EXCHANGE OF ONE NEUTRINO

Let us see whether we can get a force anything like gravitation by the exchange of one neutrino. These trial theories that we discuss are sloppily formulated and incompletely investigated, because they do not seem

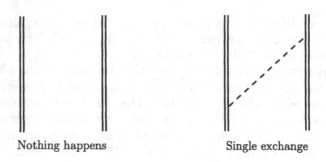

Nothing happens   Single exchange

Figure 2.3

promising when we make the first few estimates. It may be possible to patch up the difficulties that make us reject them, but I feel it is preferable to stick to the rules we have agreed upon, that we are to attempt an explanation in terms of the *known* properties of particles, without any new postulates. In that I have not succeeded.

The neutrino exchange might yield a potential which goes as $1/r$, because its mass is zero. But since the neutrino has a half-integral spin, the single exchange does not result in a static force, because after a single exchange the neutrino source is no longer in the same state it was initially. In order to get a force out of an exchange, and not only scattering, it is necessary that the diagram including the exchange should be able to interfere with the diagonal terms in the scattering amplitude, that is, it should add to the amplitude that nothing happens (Figure 2.3). Thus the possibility of one neutrino exchange is ruled out by the fact that a half unit of angular momentum cannot be emitted by an object that remains in the same internal state as it started in.

A spinless meson of zero mass would lead to a potential proportional to $1/r$. It is easiest to calculate the extra energy of the interaction in coordinate space rather than in momentum space. The momentum propagator for such a particle would be $1/k^2$. The neutrino propagator is rather $1/\rlap{/}k$, or $\rlap{/}k/k^2$, so it contains a higher power of $\vec{k}$ in the numerator; it would be even harder to reconcile such a propagator with a gravitation-like force. For the spin zero meson we calculate the diagram by integrating over all the possible emission times, and all possible capture times, the propagator $\delta_+ = 1/(t^2 - r^2 + i\epsilon)$ (Figure 2.4). The extra energy is, except for factors

$$E \propto \int \frac{idt}{(t^2 - r^2 + i\epsilon)} \quad \propto \quad \frac{1}{r}. \tag{2.3.1}$$

There is a single integration over the time of emission only (or the time of capture only); the second integration introduces a time factor which

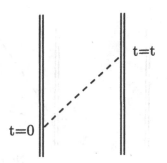

Figure 2.4

represents the normal advance of the phase, so it does not represent an interaction energy.

   We have not explicitly stated just how we expect the masses to appear as factors, but there is little point of doing this, since we do not know of any spinless neutrinos. This theory is killed by the half unit of angular momentum carried away.

2.4   THE EXCHANGE OF TWO NEUTRINOS

Maybe we can still get a theory of gravitation from neutrinos by exchanging two at a time, so that they can have diagonal expectation values. There is no obvious way in which to see why the energy of interaction between two large objects would be exactly proportional to their masses, although it is evident that it will be at least roughly proportional to the number of particles in each. Leaving this aside (to come back to if everything else worked, which it won't) we shall say the interaction of two objects is proportional to $m_1 \cdot m_2$ times the interaction of one pair of constituent particles. We proceed much as before but do it a little more carefully, since the results are more interesting. The amplitude to emit a pair of neutrinos per dt is $G'\,dt$. The amplitude that one neutrino makes its way from one point to another is $1/(t^2 - r^2 + i\epsilon)$. We introduce the masses of the interacting particles m_1, m_2 by saying that these masses are to represent the total number of particles, so that the energy between two masses is

$$E = m_1 m_2 G'^2 \int \frac{i\,dt}{(t^2 - r^2 + i\epsilon)^2}. \qquad (2.4.1)$$

This integral can be done very easily, either by poles or by simple differentiation of the single-neutrino integral (2.3.1), so that the energy is

$$E = m_1 m_2 \frac{G'^2 \pi}{2} \frac{1}{r^3}, \qquad (2.4.2)$$

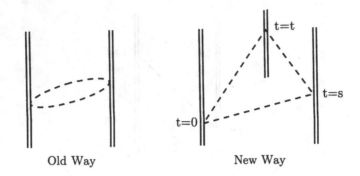

Old Way   New Way

Figure 2.5

where r is the separation of the particles. Thus we find that the two-neutrino exchange gives an energy with the wrong radial dependence. This makes the theory look hopeless. But hope arises again if we analyze still further. It turns out that we can get a term that goes as $1/r$ by considering an exchange between three masses. Three particles can exchange two neutrinos between any of the three pairs, and in a new way (Figure 2.5). Let the first emission occur at $t = 0$, and the other vertices occur at times t and s. The energy would then involve

$$G'^3 \, m_1 m_2 m_3 \, i^2 \int \frac{ds \, dt}{(s^2 - r_{12})(t^2 - r_{23})[(s-t)^2 - r_{31}]}. \qquad (2.4.3)$$

This integral may be carried out by successive integrations, each done by poles, and the answer is

$$E = -G'^3 \, m_1 m_2 m_3 \, \pi^2 \, \frac{1}{(r_{12} + r_{23} + r_{13})r_{12}r_{23}r_{13}}. \qquad (2.4.4)$$

If one of the masses, say mass 3, is far away so that r_{13} is much larger than r_{12}, we do get that the interaction between masses 1 and 2 is inversely proportional to r_{12}.

What is this mass m_3? It evidently will be some effective average over all other masses in the universe. The effect of faraway masses spherically distributed about masses 1 and 2 would appear as an integral over an average density; we would have

$$E = -\frac{G'^3 \, m_1 m_2 \, \pi^2}{r_{12}} \int \frac{4\pi \rho(R) R^2 dR}{2R^3}, \qquad (2.4.5)$$

where R is the large radius $R \approx r_{13} \approx r_{23}$. For a simple estimate, we may take the density to be constant inside a sphere; we carry out the

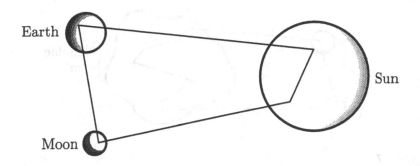

Figure 2.6

integration from an initial inside radius large enough that r_{12} is small in comparison. The contribution of all masses outside this minimum radius is something like

$$E = -\frac{m_1 m_2}{r_{12}} 2\pi^3 G'^3 \ln\left(\frac{R_0}{R_i}\right) \rho. \qquad (2.4.6)$$

This logarithm is some number which can't be much over 50 or 100, since the typical outer radius involved might be $Tc = 10^{10}$ light-years $= 10^{28}$ cm. Such an energy acts very much like gravitation; can we disprove it? Yes, in two ways. First, the size of this logarithmic term begins to be comparable to the direct force (2.4.2) proportional to $1/r^3$ at distances larger than those where the Newton law has already been verified. Moreover, if we consider the effect of the sun on the gravitational attraction between the earth and the moon, we find that it would produce observable deviations in the orbital motion of the moon, since the earth's distance from the sun varies along the orbit. We estimate this as follows. We want to compare the contribution of the sun to the earth-moon attraction with the contribution of all the other stars. The effects vary as the mass, and inversely as the cube of the distance. For a logarithm smaller than 1000, the sun contribution is 10^{12} times the star contribution, for any reasonable estimate of the star density! We may thus neglect the star contribution. But no perturbations as large as the change in the effective gravitational constant that would arise from the ±2 percent variation of the earth-sun distance have been observed in the earth-moon system.

Is it still possible to save this theory? It is conceivable that processes of higher order may remove these difficulties, for example, a process involving 4 neutrino lines or even higher orders, may be calculated. It is clear that a term like Figure 2.6 would make a contribution of order m_3^2, the square of the number of particles in the ambient masses and therefore the influence of distant nebulae would far outweigh that from the sun. This would be even more true for higher orders. We must therefore sum

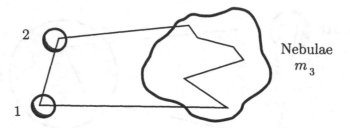

Figure 2.7

diagrams of various orders of the type in Figure 2.7. However, as far as I can see from the Fermi Statistics, the terms of various orders seem to alternate in sign and no satisfactory result is obtained.

To explain the idea more formally, what we are guessing is that in a theory with (say a scalar ϕ) quadratic coupling like $\bar{\psi}\phi^2\psi$ the expectation for the product of two field variables $\phi(1)\phi(2)$ in a state with many nebulae present may be

$$\langle\phi(1)\phi(2)\rangle = \delta_+(S_{12}^2) + C,$$

where the first term is the usual vacuum term, and C comes from interaction with distant nebulae. This C is practically independent of 1 and 2, if they are close compared to the radius of the universe.

However, for spin 1/2 I think C cannot arise. For photons such a term, even if it arose, would not have the desired effect as far as I can see, for the coupling of photons is not quadratic.

Thus, the neutrino pair theory has not led to a fruitful result. However, we have learned something from working through it, which is a very fascinating idea. If we construct a theory of gravitation based on three-body interactions, in which the body is far removed from the other two, a $1/r$ law will result if the field quanta have integral spin, and the strength of the potential will be proportional to the amount of matter in the distant body. The nebulae, which so far have not affected our physical laws, may play an important role in the gravitational interaction!

However, in order to construct such a theory, we would have to assume the existence of a boson of zero mass which coupled quadratically with all matter in the universe, and we would have to return to the question of why the force is exactly proportional to the mass. But there does not seem to be much reason to follow such a tortuous path for we have not succeeded with known particles. If we allow a new particle we can construct a perfectly good theory by assuming the existence of a spin-two particle of zero mass, which couples linearly with the matter.

3.1 THE SPIN OF THE GRAVITON

We have discussed, from the viewpoint of imaginary Venutian scientists, possible interpretations of gravity in terms of known fields. We assume that these scientists know the general properties of field theories; they are now searching for a field having the characteristics of gravity. In order to make distinctions between various possibilities, we shall need to recall the following properties of gravity; that large masses attract with a force proportional to the inertia, and to the inverse square of the distance; also that the mass and inertia represent the energy content, since the binding energies of atoms and nuclei have a gravitational behavior identical to the rest energies.

We can imagine that one group of field theorists has attempted to interpret gravity in terms of known particles, as we did in the last lecture, and has failed. Another group of field theorists has begun to deduce some of the properties of a new field that might behave as gravity.

In the first place, the fact that gravity is found to have a long range automatically means that the interaction energy depends on separation as $1/r$. There is no other possibility in a field theory. The field is carried by the exchange of a particle, which henceforth we shall call the graviton. It must have a mass $m = 0$ so that the force proportional to $1/r^2$

results from the interaction. The next guess we must make before we can write down a field theory is, what is the spin of the graviton? If the spin were 1/2, or half integral, we would run into the difficulties discussed in the last lecture (2.3), in that there could be no interference between the amplitudes of a single exchange, and no exchange. Thus, the spin of the graviton must be integral, some number in the sequence 0, 1, 2, 3, 4, ... Any of these spins would give an interaction proportional to $1/r$, since any radial dependence is determined exclusively by the mass. In order to distinguish between the various spins, we must look at subtler differences between the effects due to gravitons of the various spins. We can imagine that our group of field theorists divides up the work, so that some deduce the details of a spin- zero theory, others work on spin 1, still others on spin 2, 3, or even 4. The labor of working out the details for the higher spins is considerably greater than for the lower spins, so we take them in increasing sequence.

A spin-1 theory would be essentially the same as electrodynamics. There is nothing to forbid the existence of two spin-1 fields, but gravity can't be one of them, because one consequence of the spin 1 is that likes repel, and unlikes attract. This is in fact a property of all odd-spin theories; conversely, it is also found that even spins lead to attractive forces, so that we need to consider only spins 0 and 2, and perhaps 4 if 2 fails; there is no need to work out the more complicated theories until the simpler ones are found inadequate.

The rejection of spin-zero theories of gravitation is made on the basis of the gravitational behavior of the binding energies. We need not work out the complete details here; we shall give an argument by analogy, and then proceed directly to the construction of a spin-2 theory. We ask the question, what is the attraction between moving objects; is it larger or smaller than for static objects? We may for example calculate the mutual attraction of the two masses of gas; the experimental evidence on gravity suggests that the force is greater if the gases are hotter (Figure 3.1).

We know what happens in electrodynamics. The electric forces are unchanged by random motions of the particles. Now the interaction energy is proportional to the expectation value of the operator γ_t, which is $1/\sqrt{1 - v^2/c^2}$. Since the potential resulting from this operator is not velocity dependent the proportionality factor must go as $\sqrt{1 - v^2/c^2}$. This means that the interaction energy resulting from the operator 1, corresponding to the spin-0 field, would be proportional to $\sqrt{1 - v^2/c^2}$. In other words, the spin-zero theory would predict that the attraction between masses of hot gas would be *smaller* than for cool gas. In a similar way, it can be shown that the spin-2 theory leads to an interaction energy which has $\sqrt{1 - v^2/c^2}$ in the denominator, in agreement with the experimental results on the gravitational effect of binding energies. Thus,

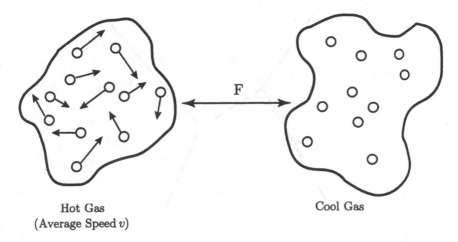

Hot Gas
(Average Speed v)

Cool Gas

Figure 3.1

the spin-0 theory is out, and we need spin 2 in order to have a theory in which the attraction will be proportional to the energy content.

3.2 AMPLITUDES AND POLARIZATIONS IN ELECTRODYNAMICS, OUR TYPICAL FIELD THEORY

Our program is now to construct a spin-2 theory in analogy to the other field theories that we have. We could at this point switch to Einstein's viewpoint on gravitation, since he obtained the correct theory, but it will be instructive and possibly easier for us to learn if we maintain the fiction of the Venutian scientists in order to guess at the properties of the correct theory. In thus implying that many scientists *today* could arrive at the correct theory of gravitation, we are not detracting from Einstein's achievement. Today we have a hindsight, a developed formalism, which did not exist fifty years ago, and also we have Einstein pointing out the direction. It is very difficult to imagine what we would do if we didn't know what we know, when we know it, but let us proceed and guess at the correct theory in analogy with electrodynamics.

In the theories of scalar, vector, and tensor fields (another way of denoting spins 0, 1, and 2) the fields are described by scalar, vector, or tensor potential functions:

Spin 0	X	Scalar Potential
Spin 1	A_μ	Vector Potential
Spin 2	$h_{\mu\nu}$	Symmetric Tensor Potential

Another theory would result from assuming that the tensor is antisymmetric; it would not lead to something resembling gravity, but rather

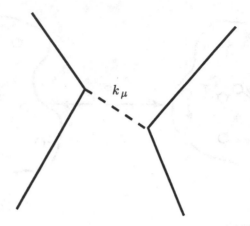

Figure 3.2

something resembling electromagnetism; the six independent components of the antisymmetric tensor would appear as two space vectors.

The source of electromagnetism is the vector current j_μ, which is related to the vector potential by the equation

$$A_\mu = -\frac{1}{k^2}\, j_\mu. \tag{3.2.1}$$

Here we have taken Fourier transforms and use the momentum-space representation. The D'Alembertian operator is simply k^2 in momentum space. The calculation of amplitudes in electromagnetism is made with the help of propagators connecting currents in the manner symbolized by diagrams such as that in Figure 3.2. We compute amplitudes for such processes as a function of the relativistic invariants, and restrict our answer as demanded by rules of momentum and energy conservation. The guts of electromagnetism are contained in the specification of the interaction between a current and the field as $j^\mu A_\mu$; in terms of the sources, this becomes an interaction between two currents:

$$-j'_\mu \frac{1}{k^2}\, j^\mu. \tag{3.2.2}$$

For a particular choice of coordinate axes, the vector k_μ may be expressed as

$$k^\mu = (\omega, \kappa, 0, 0). \tag{3.2.3}$$

Note that we use an index ordering 4, 3, 2, 1 such that

$$\begin{aligned} x^\mu &= (t, z, y, x) \\ A_\mu &= (A_4, A_3, A_2, A_1). \end{aligned} \tag{3.2.4}$$

Then the current-current interaction when the exchanged particle has a four momentum k^μ is given by

$$-j'_\mu \left(\frac{1}{k^2}\right) j^\mu = -\frac{1}{\omega^2 - \kappa^2} \left(j'_4 j_4 - j'_3 j_3 - j'_2 j_2 - j'_1 j_1\right). \qquad (3.2.5)$$

The conservation of charge, which states that the four divergence of the current is zero, in momentum space becomes simply the restriction

$$k_\mu j^\mu = 0. \qquad (3.2.6)$$

In the particular coordinate system we have chosen, this restriction connects the third and fourth component of the currents by

$$\omega j^4 - \kappa j^3 = 0, \qquad \text{or} \qquad j^3 = \frac{\omega}{\kappa} j^4. \qquad (3.2.7)$$

If we insert this expression for j_3 into the amplitude (3.2.5), we find that

$$-j'_\mu \left(\frac{1}{k^2}\right) j^\mu = \frac{j'_4 j_4}{\kappa^2} + \frac{1}{\omega^2 - \kappa^2} \left(j'_1 j_1 + j'_2 j_2\right). \qquad (3.2.8)$$

Now we may give an interpretation to the two terms of this equation. The fourth component of the current is simply the charge density; in the situation that we have stationary charges it is the only non-zero component. The first term is independent of the frequency; when we take the inverse Fourier transform in order to convert this to a space-interaction, we find that it represents an instantaneously acting Coulomb potential.

$$(\text{F.T.})^{-1} \left[\frac{j'_4 j_4}{\kappa^2}\right] = \frac{e^2}{4\pi r} \delta(t - t') \qquad (3.2.9)$$

This is always the leading term in the limit of small velocities. The term appears instantaneous, but this is only because the separation we have made into two terms is not manifestly covariant. The total interaction is indeed a covariant quantity; the second term represents corrections to the instantaneous Coulomb interaction.

The interaction between two currents always involves virtual photons. We can learn something about the properties of real photons by looking at the poles of the interaction amplitude, which occur for $\omega = \pm \kappa$. Of course, any photon that has a physical effect may be considered as a virtual photon, since it is not observed unless it interacts, so that observed photons never really have $\omega = \pm \kappa$. There are however no difficulties in passing to the limit; physically, we know of photons that come from the moon, or the sun, for which the fractional difference between ω and κ is very, very small. If we consider that we also observe photons from distant galaxies which are millions of light years away, we see that it must make physical sense to think we are so close to the pole that for these there can't be any physical effect *not* like the pole term. The residue of the pole term at $\omega = \kappa$ is the sum of two terms each of which is the product of two factors. It seems that there is one kind of photon which interacts with j_1 and j_1', and another kind of photon that interacts with j_2 and j_2'. In the usual language, we describe this by saying that there are two independent polarizations for photons.

The circular polarizations are nothing but linear combinations of plane polarized photons, corresponding to a separation of the sum of products $(j_1' j_1 + j_2' j_2)$ in a different basis; we have

$$
\begin{aligned}
\left(j_1' j_1 + j_2' j_2\right) = \; & \frac{1}{\sqrt{2}} \left(j_1 + i j_2\right) \frac{1}{\sqrt{2}} \left(j_1' + i j_2'\right)^* \\
& + \frac{1}{\sqrt{2}} \left(j_1 - i j_2\right) \frac{1}{\sqrt{2}} \left(j_1' - i j_2'\right)^* . \quad (3.2.10)
\end{aligned}
$$

Still we see that there are two kinds of photons. The advantage of this separation becomes apparent when we consider a rotation of the coordinate system about the direction of propagation 3. The circularly polarized photons rotate into themselves, so to speak, because they change only by a phase as the coordinates are rotated by an angle θ; the phases are $\exp(i\theta)$ and $\exp(-i\theta)$.

The quantum-mechanical rules describing the behavior of systems under rotations tell us that systems having this property are in a state of unique angular momentum; the photons that change in phase as $\exp(i\theta)$ have an angular momentum projection 1, and the others a projection -1.

We might expect that if the photons are objects of spin 1, there might be a third kind of photon having a spin projection 0. However, it can be shown, that for a relativistic theory of particles with zero rest mass, only two projection states are allowed, having the maximum and minimum values of projection along the direction of propagation. This is a general result, good for particles with any spin, which has been proved by Wigner. We shall not carry out the proof here, but simply in the cases of interest, the photon now and the graviton later, we will show the existence of these two states by explicit separation of the interaction.

It may be objected that we have shown the existence of these two states for the *currents*, which are source operators, rather than for the photons themselves. But one implies the other, since the amplitude to emit a photon of circular polarization -1 is given by the current operator $(j_1 + i j_2)/\sqrt{2}$, etc. As we rotate the coordinate system, the amplitude for emission should not change, so thus we arrive at the requirement of appropriate phase changes for the photons themselves. The actual polarization of a photon is perhaps best defined in terms of the projections of the vector potential in specific directions, such as e_μ (e_μ is a unit vector). The interaction of such a photon with a current j^μ, i.e., the amplitude to absorb or emit such a photon, is given by

$$-e_\mu\, j^\mu = (\text{projection of } j^\mu \text{ along } e_\mu). \qquad (3.2.11)$$

3.3 AMPLITUDES FOR EXCHANGE OF A GRAVITON

We shall write down the amplitudes for the exchange of a graviton by simple analogy to electrodynamics. We shall have to pay particular attention to the instantaneous, nonrelativistic terms, since only these are apparent in the present experimental observations of gravity. The full theory gives us both the instantaneous terms (analogous to the Coulomb interaction), and the corrections which appear as retarded waves; we will have to separate out these retardation effects for calculations of observable effects.

We assume the D'Alembertian in momentum space is k^2; by simple analogy with equation (3.2.1) we expect the field tensor to be related to its source tensor as follows

$$h_{\mu\nu} = \frac{1}{k^2}\, T_{\mu\nu}. \qquad (3.3.1)$$

What can the interaction be? Since electrodynamics relates currents, let us guess that the source tensors appear in the interaction energy as

$$T'_{\mu\nu}\left(\frac{1}{k^2}\right) T^{\mu\nu}. \qquad (3.3.2)$$

It is our job now to ascribe particular characteristics to the tensor **T** so that the characteristics of gravity are reproduced. It is possible *a priori* that the tensor **T** involves gradients, that is, the vector $\vec{k}$. If only gradients are involved, the resulting theory has no monopoles; the simplest objects would be dipoles. We want the tensor **T** to be such that in the nonrelativistic limit, the energy densities appear in analogy to the charge densities j_4. As is well known, we have in electromagnetism a stress tensor whose component T_{44} is precisely the energy density of the electromagnetic field. It is therefore quite likely that there is some general tensor whose component T_{44} is the total energy density; this will give the Newtonian law of gravity in the limit of small velocities, an interaction energy

$$-\frac{T'_{44}\, T_{44}}{\kappa^2}. \qquad (3.3.3)$$

Then, in order to have a correct relativistic theory, it must follow that the amplitude involves the complete tensor **T**, as we have guessed in (3.3.2).

There is one point which we haven't mentioned about this tensor. The trace of a symmetric tensor is an invariant quantity, not necessarily zero. Thus in computing from a symmetric tensor of nonzero trace we might get a theory which is a mixture of spin 0 and spin 2. If we write a theory using such a tensor, we will find, when we come to separating the interaction into its polarizations, that there are apparently three polarizations instead of the two allowed to a massless spin-2 particle. To be more explicit, we may have besides the interaction (3.3.2) another possible invariant form proportional to $T^\mu{}_\mu (1/k^2) T^\nu{}_\nu$. We shall try to adjust the proportions of these two so that no real gravitons of angular momentum zero would be exchanged.

We write explicitly the various terms as follows:

$$T'_{\mu\nu} \left(\frac{1}{k^2} \right) T^{\mu\nu} = \frac{1}{\omega^2 - \kappa^2} \Big[T'_{44} T_{44} - 2T'_{43} T_{43} - 2T'_{42} T_{42} - 2T'_{41} T_{41}$$

$$+ 2T'_{23} T_{23} + 2T'_{31} T_{31} + 2T'_{21} T_{21} + T'_{33} T_{33} + T'_{22} T_{22} + T'_{11} T_{11} \Big]. \quad (3.3.4)$$

In electrodynamics, we had obtained a simplification by using the law of conservation of charge. Here, we obtain a simplification by using the law of energy conservation, which is expressed in momentum space by

$$k^\mu T_{\mu\nu} = 0. \quad (3.3.5)$$

In our usual coordinate system, such that k^1 and k^2 are zero, this relates the 4 index to the 3 index of our tensor by

$$\omega T_{4\nu} = -\kappa T_{3\nu}. \quad (3.3.6)$$

When we use this relation to eliminate the index 3, we find that the amplitude separates into an instantaneous part, having a typical denominator κ^2, and a retarded part, with a denominator $(\omega^2 - \kappa^2)$. For the instantaneous term we get

$$-\frac{1}{\kappa^2} \left[T'_{44} T_{44} \left(1 - \frac{\omega^2}{\kappa^2} \right) - 2T'_{41} T_{41} - 2T'_{42} T_{42} \right], \quad (3.3.7)$$

and for the retarded term

$$\frac{1}{\omega^2 - \kappa^2} \left[T'_{11} T_{11} + T'_{22} T_{22} + 2T'_{21} T_{21} \right]. \quad (3.3.8)$$

The transverse components of the tensor **T** are presumably independent, so that this represents a sum of three independent products, or three polarizations. We see that this theory contains a mixture of spin 0 and spin 2. To eliminate the spin zero part we must add to our amplitude a term of the form

$$\alpha \, T^{\prime\nu}{}_{\nu} \left(\frac{1}{k^2} \right) T^{\mu}{}_{\mu}. \tag{3.3.9}$$

In the retarded term, this adds pieces of the tensor as follows:

$$\alpha \, \frac{1}{\omega^2 - \kappa^2} \left(T^{\prime}_{11} + T^{\prime}_{22} \right) \left(T_{11} + T_{22} \right).$$

We can adjust the parameter α so that the retarded term contains only a sum of two independent products. The proper value of α is $-1/2$, to make the retarded term equal to

$$\frac{1}{\omega^2 - \kappa^2} \left[\frac{1}{2} \left(T^{\prime}_{11} - T^{\prime}_{22} \right) \left(T_{11} - T_{22} \right) + 2 T^{\prime}_{12} T_{12} \right]. \tag{3.3.10}$$

There are then two directions of the polarization, which are generated by these combinations of the tensor elements

$$\frac{1}{\sqrt{2}} \left(T_{11} - T_{22} \right) \quad \text{and} \quad \sqrt{2} \left(T_{12} \right). \tag{3.3.11}$$

The different normalization is a result of the symmetry of our tensor; we can restore some of the symmetry by writing

$$\sqrt{2} \, T_{12} = \frac{1}{\sqrt{2}} \left(T_{12} + T_{21} \right). \tag{3.3.11a}$$

A possible plane-wave solution representing our graviton is therefore

$$h_{\mu\nu} = e_{\mu\nu} \exp \left(i k_{\sigma} x^{\sigma} \right), \tag{3.3.12}$$

where the polarization tensor $e_{\mu\nu}$ has the following nonzero components:

$$e_{11} = \frac{1}{\sqrt{2}}, \quad e_{22} = -\frac{1}{\sqrt{2}}, \quad \text{and} \quad e_{12} = e_{21} = \frac{1}{\sqrt{2}}. \tag{3.3.13}$$

Our interaction in general,

$$T^{\prime}_{\mu\nu} \frac{1}{k^2} T^{\mu\nu} - \frac{1}{2} T^{\prime\mu}{}_{\mu} \frac{1}{k^2} T^{\nu}{}_{\nu},$$

can be written as

$$T^{\prime\sigma\tau} P_{\sigma\tau,\mu\nu} T^{\mu\nu},$$

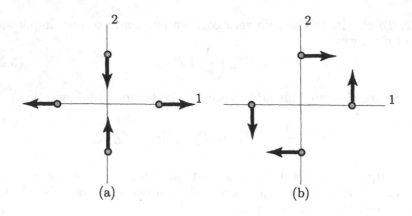

Figure 3.3

where $P_{\sigma\tau,\mu\nu}$, the propagator for a graviton, is

$$\frac{1}{2}\left(\eta_{\mu\sigma}\eta_{\nu\tau} + \eta_{\mu\tau}\eta_{\nu\sigma} - \eta_{\mu\nu}\eta_{\sigma\tau}\right)\frac{1}{k^2}.$$

We shall usually prefer for simplicity to keep the propagator as a simple factor $1/k^2$ and to represent the interaction by virtual gravitons, emitted from the source with amplitude

$$h_{\mu\nu} = \frac{1}{k^2}\left(T_{\mu\nu} - \frac{1}{2}\eta_{\mu\nu}T^\sigma{}_\sigma\right)$$

and with a coupling $h_{\mu\nu}T'^{\mu\nu}$ for absorption.

The amplitude for emission of a real graviton of polarization $e_{\sigma\tau}$ if $e^\sigma{}_\sigma = 0$ as in (3.3.13), is given by the inner product $e_{\sigma\tau}T^{\sigma\tau}$.

3.4 PHYSICAL INTERPRETATION OF THE TERMS IN THE AMPLITUDES

The polarization of a graviton is a tensor quantity. We may visualize this with pictures similar to those we use in describing stresses; we draw arrows indicating the direction to be associated with surfaces normal to the axes. In the plane perpendicular to the direction of propagation we have the two stresses in Figure 3.3. These are the only two possible quadrupole stresses; the stresses representable by all arrows pointing towards the origin (or away from the origin) are something like a fluid pressure, which has zero spin. The "stresses" (actually rotations) representable by all arrows pointing in a clockwise (or counter-clockwise) direction correspond to spin 1.

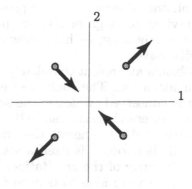

Figure 3.4

The stresses represented by Figure 3.3 (b) may be referred to axes which are 45° from the original axes; in this case the picture is Figure 3.4 which is nothing but the same stress of Figure 3.3 (a), rotated by 45°. From this we find that these two polarizations turn into each other with a rotation of axes by 45°. As we rotate by 90°, each polarization returns to itself; the arrows are reversed, but we must think of an oscillating time dependence associated with the polarizations. Continuing in this way, we see a complete 360° rotation corresponds to two complete cycles of phase— the spin is two. There are two orthogonal linear combinations of these two polarizations, whose rotational phase change behaves as $\exp(2i\theta)$ and $\exp(-2i\theta)$. This is simply a different separation of the retarded term; with a little trial and error we can simply display the two parts

$$\frac{1}{4} \left(T'_{11} - T'_{22} + i2T'_{12}\right) \left(T_{11} - T_{22} - i2T_{12}\right)$$

$$+ \frac{1}{4} \left(T'_{11} - T'_{22} - i2T'_{12}\right) \left(T_{11} - T_{22} + i2T_{12}\right). \quad (3.4.1)$$

That these have the character of spin 2, projection ±2 tensors is obvious when we compare the form of these products to the product of harmonic polynomials; we know that $(x \pm iy)(x \pm iy)$ are evidently of spin 2 and projection ±2; these products are $(xx - yy \pm 2ixy)$, which have the same structure as our terms (3.4.1). Thus we conclude that with the choice $\alpha = -1/2$, our gravitons have only two possible polarizations. This is possibly the correct theory, equivalent to a field theory of spin 2 which our field theorists Pauli and Fierz [FiPa 39] have already worked out and expressed in terms of field Lagrangians.

We are approaching the spin-2 theory from analogies to a spin-1 theory; thus we have without explanation assumed the existence of graviton plane waves, since the photon plane waves are represented by poles of the propagator, and the graviton propagator also has poles at $\omega = \pm \kappa$. But the experimental evidence is lacking; we have observed neither gravitons, nor even classical gravity waves.

There are some problems at present completely disregarded to which we shall later turn our attention. The sources of electromagnetism are conserved, and energy is conserved, which is the source of gravity. But this conservation is of a different character, since the photon is uncharged, and hence is not a source of itself, whereas the graviton has an energy content equal to $\hbar \omega$, and therefore it is itself a source of gravitons. We speak of this as the nonlinearity of the gravitational field.

In electromagnetism, we were able to deduce field equations (Maxwell's equations), which are inconsistent if charge is not conserved. We have so far avoided discussion of a field equation for gravity by worrying only about the amplitudes, and not the field themselves. Also, we need yet to discuss whether the theory we can write will be dependent on a gauge, and whether we can at all write a field equation corresponding to Maxwell's $\partial F^{\mu\nu}/\partial x^\nu = j^\mu$.

There are some physical consequences of our theory that may be discussed without field equations, by simply considering the form of the interaction. We write the complete expression corresponding to $\alpha = -1/2$:

$$2\left[T'_{\mu\nu} \left(\frac{1}{k^2} \right) T^{\mu\nu} - \frac{1}{2} T'^\nu{}_\nu \left(\frac{1}{k^2} \right) T^\mu{}_\mu \right]$$

$$= -\frac{1}{\kappa^2} \left[T'_{44} T_{44} \left(1 - \frac{\omega^2}{\kappa^2} \right) + T_{44} \left(T'_{11} + T'_{22} \right) \right.$$

$$\left. + T'_{44} \left(T_{11} + T_{22} \right) - 4 T'_{41} T_{41} - 4 T'_{42} T_{42} \right]$$

$$- \frac{1}{\kappa^2 - \omega^2} \left[\left(T'_{11} - T'_{22} \right) \left(T_{11} - T_{22} \right) + 4 T'_{12} T_{12} \right].$$

(If desired, the term $(\omega^2/\kappa^2) T'_{44} T_{44}$ may be replaced by $T'_{43} T_{43}$ or by $(T_{44} T'_{33} + T_{33} T'_{44})$.) We have already discussed the retarded term and its polarizations. Let us now look at the first term. The tensor $\mathbf{T}$ is the stress density; for slow particles, the space components are of order v/c, so that the Newtonian law is represented by only one of the products, $T'_{44} T_{44}$. The other products are somewhat analogous to magnetism. Note that in this separation they appear as instantaneous terms. The retardation effects, the travelling waves, appear only at even powers of v/c.

We might think that the instantaneous magnetic-like terms would have observable effects, for example, there might be a small change in the gravitational attraction between two wheels if we spin them faster and faster. The theory indeed predicts this, but in practice these forces would not only be very, very small, but also they would be masked by a multitude of other effects. Magnetic forces, such as the attraction between two current-carrying wires, are conveniently observable only because the direct Coulomb effects are cancelled out very, very exactly by the presence of equal numbers of positive and negative charges. But all gravitational forces are attractive, so there is no hope of such cancellation. For spinning wheels, there would be trouble in that the elastic stresses of matter would add energy terms, the wheels would fly apart, etc. On top of this, we might think that the ordinary gravitational attraction is hard enough to measure, and that the magnetic-like effects might be smaller by some ratio $(v/c)^2$ such as the ratio of magnetic to Coulomb forces. Forces between current-carrying wires are of the order of a gram's weight, whereas the uncancelled Coulomb attractions between the particles in the wires are of the order of billions of millions of tons.

It may be possible to observe effects due to this magnetic-like term if we consider the gravitational attraction of particles moving at the speed of light, or very near it. Suppose $T'_{\mu\nu}$ is due to a stationary source, such as the sun, so that only T'_{44} survives, and we consider the gravitational attraction between the sun and a fast particle, which is travelling with a velocity v near c, so that its stress tensor has components such that $T_{11} = (v^2/c^2)T_{44}$. Then in (3.4.2) we see that the interaction energy is larger than that due to T_{44} alone, by a factor of $1 + v^2/c^2$, or 2 for a photon. Thus, as a photon travels in a strong gravitational field, it falls by an amount larger than the Newtonian theory would predict from its energy content alone. The deflections of the light of a star as it passes near the surface of the sun are twice as great as an "impulse" calculation would give in Newtonian theory. The terrestrians have done this experiment, and they do find that the deflection is greater than given by the falling theory, by a factor very close to 2. Even though the experimental evidence is rather imperfect and is not all consistent, the evidence strongly favors a real effect in the direction that our theory predicts.

We might at this point proceed to calculate in detail effects such as this, and also many other problems, such as Compton scattering of gravitons, effects on the motion of the planet Mercury about the sun, in order to find out the orders of magnitudes of gravitational effects, and what experiments might be possible. However, it is perhaps preferable to proceed to a description of the gravitational field itself, in terms of the field Lagrangian, and field equations, rather than the amplitudes.

3.5 THE LAGRANGIAN FOR THE GRAVITATIONAL FIELD

We shall now study our theory in terms of a Lagrangian, studying the fields themselves rather than simply the amplitudes. We first review the situation in electrodynamics. There, the action is

$$S_{\mathrm{E}} = - \int d\tau \left[\frac{1}{4} \left(\frac{\partial A_\mu}{\partial x^\nu} - \frac{\partial A_\nu}{\partial x^\mu} \right) \left(\frac{\partial A^\mu}{\partial x_\nu} - \frac{\partial A^\nu}{\partial x_\mu} \right) + j^\mu A_\mu \right]. \quad (3.5.1)$$

It is from such a Lagrangian that we eventually deduce field equations; we want the gravitational analogue of $A_\mu = -(1/k^2)j_\mu$.

It is not difficult to guess at the form of the second, or coupling, term. This we guess to be $-\lambda h_{\mu\nu}T^{\mu\nu}$. Now the analogy for the terms involving the derivatives of $h_{\mu\nu}$ is not so obvious; there are simply too many indices which can be permuted in too many ways. We will have to write a general form for the Lagrangian, as a sum over all the ways of writing the field derivatives, putting arbitrary coefficients in front of each term, as follows:

$$a \left(\frac{\partial h^{\mu\nu}}{\partial x_\sigma} \cdot \frac{\partial h_{\mu\nu}}{\partial x^\sigma} \right) + b \left(\frac{\partial h^{\mu\sigma}}{\partial x_\nu} \cdot \frac{\partial h_{\mu\nu}}{\partial x^\sigma} \right) + c \left(\frac{\partial h^\mu{}_\mu}{\partial x_\nu} \cdot \frac{\partial h^\sigma{}_\nu}{\partial x^\sigma} \right) + \ldots.$$
$$(3.5.2)$$

Our theory will not be complete until we have invented some criterion for assigning values to the coefficients $a, b, c, d, e, \ldots$.

Perhaps we can make a guess by some analogy to electromagnetism. If we vary the general Lagrangian in eq.(3.5.1), with respect to A, we arrive at a differential equation connecting the fields and the current,

$$\frac{\partial}{\partial x_\nu} \frac{\partial}{\partial x^\nu} A_\mu - \frac{\partial}{\partial x_\nu} \frac{\partial}{\partial x^\mu} A_\nu = j_\mu. \quad (3.5.3)$$

For economy of writing, we shall henceforth indicate such differentiations (gradients) by simply indicating the index of the coordinates after a comma; the equation above becomes

$$A_{\mu,\nu}{}^\nu - A_{\nu,\mu}{}^\nu = j_\mu. \quad (3.5.4)$$

The conservation of charge is expressed by taking the divergence of j_μ equal to zero. But we may notice that the Maxwell equations for the field are not consistent unless there is charge conservation, and that the gradient of the expression on the left hand side of (3.5.4) is *identically* zero. With the correct electromagnetic Lagrangian, then, the conservation of charge can be deduced as a consequence of the field equations. The *left* hand side satisfies an *identity*, its divergence is zero:

$$A_{\mu,\nu}{}^{\nu\mu} - A_{\nu,\mu}{}^{\nu\mu} = 0. \quad (3.5.5)$$

A similar requirement serves to define the size of the coefficients
$a, b, c, d, e, \ldots$ relative to each other. We will write a general Lagrangian,
deduce the differential field equations by varying it, and then demand
that since the divergence of the tensor **T** vanishes, the field quantities
that are equal to it should have a divergence which vanishes identically.
This will result in a unique assignment for the values of the coefficients.
We will carry out the algebra explicitly, adjusting the coefficients so that
the field equations are consistent only if

$$T^{\mu\nu}{}_{,\nu} = 0. \tag{3.5.6}$$

3.6 THE EQUATIONS FOR THE GRAVITATIONAL FIELD

We begin by writing down all the possible products of derivatives of our
field tensor $h_{\mu\nu}$. At each step, considerable simplification results if we use
the symmetry of $h_{\mu\nu}$ in order to combine terms. If the two tensor indices
are different from the derivative index, we have two possible products:

1. $h_{\mu\nu,\sigma}\, h^{\mu\nu,\sigma}$
2. $h_{\mu\nu,\sigma}\, h^{\mu\sigma,\nu}$

If there are two indices which are equal, we may have three possible
products:

3. $h^{\mu\nu}{}_{,\nu}\, h^{\sigma}{}_{\mu,\sigma}$
4. $h^{\mu\nu}{}_{,\nu}\, h^{\sigma}{}_{\sigma,\mu}$
5. $h^{\nu}{}_{\nu,\mu}\, h_{\sigma}{}^{\sigma,\mu}$

Not all the five products are necessary; number 2 may be omitted because
it can be converted to number 3 by integration by parts. This leaves only
four independent products of derivatives. That is, we assume an action
of the form

$$S = \int d\tau \left[a h^{\mu\nu,\sigma}\, h_{\mu\nu,\sigma} + b h^{\mu\nu}{}_{,\nu}\, h^{\sigma}{}_{\mu,\sigma} + c h^{\mu\nu}{}_{,\nu}\, h^{\sigma}{}_{\sigma,\mu} \right.$$
$$\left. + d\, h^{\nu}{}_{\nu,\mu}\, h_{\sigma}{}^{\sigma,\mu} - \lambda T^{\mu\nu} h_{\mu\nu} \right]. \tag{3.6.1}$$

Now we vary this sum of four products with respect to $h_{\alpha\beta}$ in order to
obtain a differential equation relating the field derivatives with the source
tensor $T_{\alpha\beta}$. The result is (watch out to remember that $\delta h_{\alpha\beta}$ is symmetric
in α, β, so only the symmetric part of its coefficient must be zero)

$$a\, 2h_{\alpha\beta,\sigma}{}^{,\sigma} + b\left(h_{\alpha\sigma,\beta}{}^{,\sigma} + h_{\beta\sigma,\alpha}{}^{,\sigma} \right)$$
$$+ c\left(h^{\sigma}{}_{\sigma,\alpha\beta} + \eta_{\alpha\beta}\, h^{\mu\nu}{}_{,\nu\mu} \right) + d\, 2\eta_{\alpha\beta}\, h^{\sigma}{}_{\sigma,\mu}{}^{,\mu} = -\lambda T_{\alpha\beta}. \tag{3.6.2}$$

We take the derivative of each of these with respect to the index β, then the requirement that the divergence of the left should be identically zero gives the equation

$$2a\, h^{\alpha\beta,\sigma}{}_{,\sigma\beta} + b\, h^{\alpha\sigma,\beta}{}_{,\sigma\beta} + b\, h^{\beta\sigma,\alpha}{}_{,\sigma\beta} + c\, h^{\sigma}{}_{\sigma}{}^{,\alpha\beta}{}_{,\beta}$$
$$+ c\, h^{\mu\nu}{}_{,\mu\nu}{}^{,\alpha} + 2d\, h^{\sigma}{}_{\sigma,\mu}{}^{,\mu\alpha} = 0. \quad (3.6.3)$$

We now gather terms of the same ilk, and set the coefficients equal to zero; this involves occasional flipping of indices, and interchanging index labels:

$$h^{\alpha\beta,\sigma}{}_{,\sigma\beta}\,(2a + b) = 0$$
$$h^{\beta\sigma,\alpha}{}_{,\beta\sigma}\,(b + c) = 0 \qquad\qquad (3.6.4)$$
$$h^{\sigma}{}_{\sigma,\beta}{}^{,\alpha\beta}\,(c + 2d) = 0.$$

If we choose a scale for our results, such that $a = 1/2$, we obtain

$$a = \frac{1}{2} \qquad b = -1 \qquad c = 1 \qquad d = -\frac{1}{2}. \quad (3.6.5)$$

Presumably, we have now obtained the correct Lagrangian for the gravity field. As consequences of this Lagrangian, we shall eventually get a field equation.

3.7 DEFINITION OF SYMBOLS

The manipulations of these tensor quantities becomes increasingly tedious in the work that is to follow; in order not to get bogged down in the algebra of many indices, some simplifying tricks may be developed. It is not necessarily obvious at this point that the definitions we are about to make are useful; the justification comes in our later use.

We define a "bar" operation on an arbitrary second rank tensor by

$$\overline{X}_{\mu\nu} = \frac{1}{2}\,(X_{\mu\nu} + X_{\nu\mu}) - \frac{1}{2}\,\eta_{\mu\nu}X^{\sigma}{}_{\sigma}. \quad (3.7.1)$$

For a symmetric tensor such as **h**, the rule is simpler because the two terms in the first parenthesis are equal

$$\overline{h}_{\mu\nu} = h_{\mu\nu} - \frac{1}{2}\,\eta_{\mu\nu}h^{\sigma}{}_{\sigma} \qquad\qquad (3.7.2a)$$

$$\overline{\overline{h}}_{\mu\nu} = h_{\mu\nu} \qquad\qquad (3.7.2b)$$

and notice that the bar operation is its own reciprocal for symmetric tensors.

Define also the use of the unindexed tensor symbol to represent its trace

$$h = \text{Tr}(\mathbf{h}) = h^\sigma{}_\sigma$$
$$\overline{h}^\sigma{}_\sigma = -h.$$

(3.7.3)

With these notations, the field equations (3.6.2) and (3.6.5) may be written as follows, in a symmetrized version.

$$h_{\mu\nu,\sigma}{}^{,\sigma} - 2\overline{h}_{\mu\sigma,\nu}{}^{,\sigma} = -\lambda \overline{T}_{\mu\nu}$$

(3.7.4)

To get a relation for $T_{\mu\nu}$, we simply "bar" both sides of this equation.

Next we look for something analogous to the gauge invariance properties of electrodynamics to simplify the solution of (3.7.4). In electrodynamics, the field equation

$$A^{\mu,\nu}{}_{,\nu} - A_\nu{}^{,\nu\mu} = j^\mu$$

(3.7.5)

resulted in the possibility of describing fields equally well in terms of a new four vector A'_μ, obtainable from A_μ by addition of a gradient of a scalar function X.

$$A'_\mu = A_\mu + X_{,\mu}$$

(3.7.6)

What might be the analogous property of a tensor field? We guess that the following might hold: (we have to be careful to keep our tensors symmetric!) substitution of

$$h'_{\mu\nu} = h_{\mu\nu} + X_{\mu,\nu} + X_{\nu,\mu}$$

(3.7.7)

into the left side of (3.7.4) will not alter the form of that equation. The proof of this is left as an exercise.

Using this property of gauge invariance, it will be simpler to obtain equations for the fields in a definite gauge which is more appropriate, something like the Lorentz gauge of electrodynamics. By analogy with the choice

$$A^\nu{}_{,\nu} = 0$$

(3.7.8)

we shall make the corresponding choice (which we shall call a Lorentz condition)

$$\overline{h}^{\mu\sigma}{}_{,\sigma} = 0.$$

(3.7.9)

This results in field equations relating the bar of $\mathbf{T}$ to the field,

$$h^{\mu\nu,\sigma}{}_{,\sigma} = -k^2 h^{\mu\nu} = -\lambda \overline{T}^{\mu\nu}.$$

(3.7.10)

Or solving $h_{\mu\nu} = (\lambda/k^2)\overline{T}_{\mu\nu}$. It will follow immediately that the amplitude of interaction of such an $\mathbf{h}$ with another source $T'_{\mu\nu}$ from the $\lambda h_{\mu\nu} T'^{\mu\nu}$ in the Lagrangian is

$$\lambda^2 T'_{\mu\nu} \left[\frac{1}{k^2} \right] \overline{T}^{\mu\nu}.$$

Thus, this produces precisely what we have found before in discussing the amplitudes directly.

4.1 THE CONNECTION BETWEEN THE TENSOR RANK
   AND THE SIGN OF A FIELD

We would like to deduce some useful general properties of the fields, by using the properties of the Lagrangian density. For the gravitational field, we define at this point the coupling constant and normalization of plane waves that we will use from now on. We will let

$$\lambda = \sqrt{8\pi G}. \tag{4.1.1}$$

Here, G is the usual gravitational constant in natural units ($\hbar = c = 1$); the square root is included in the definition so that the constant λ is analogous to the electron charge e of electrodynamics, rather than to the square. The factor of $\sqrt{8\pi}$ serves to eliminate irrelevant factors from the most useful formulae. To represent plane-wave gravitons, we shall use fields

$$h_{\mu\nu} = e_{\mu\nu} \exp(ik \cdot x), \tag{4.1.2}$$

with the polarization tensor $e_{\mu\nu}$ normalized in such a way that

$$e_{\mu\nu}\overline{e}^{\mu\nu} = 1. \tag{4.1.3}$$

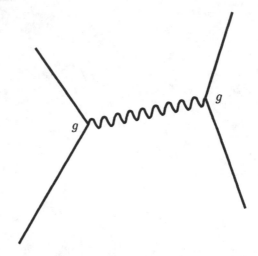

Figure 4.1

The action that describes the total system of gravity fields, matter, and coupling between matter and gravitons, has the following form:

$$S = \frac{1}{2} \int dV \left(h^{\mu\nu,\lambda} \, \overline{h}_{\mu\nu,\lambda} - 2\overline{h}^{\mu\lambda}{}_{,\lambda} \overline{h}_{\mu\nu}{}^{,\nu} \right) \quad \text{(fields)}$$

$$- \int dV \left(\lambda \, h_{\mu\nu} \, T^{\mu\nu} \right) \quad \text{(coupling term)} \quad (4.1.4)$$

$$+ S_M \quad \text{(matter)}.$$

We may from the Lagrangians of fields deduce some important properties, for example, we can understand why gravitation is attractive for likes and unlikes, whereas in electricity likes repel and unlikes attract. It can be shown that this property is inherent in the sign of the Lagrangian, so that if we change $S \to -S$, the force changes sign. The sign of the coupling constant λ or e or g makes no difference, since it appears as a square in any diagram which represents a correction to the energy; always two vertices are involved. We can change the sign of the energy corresponding to a diagram such as Figure 4.1 only if we can introduce a factor i at each vertex, for example if we are to use fields $i\phi$ rather than ϕ.

However, the fields ϕ must represent appropriate plane waves, which are consistently defined so that the standing waves in a large box have positive energies, and the quantum mechanical oscillators which represent these standing waves behave properly. Scalar fields have plane waves

$$\phi = a \exp(ik \cdot x). \quad (4.1.5)$$

The amplitude a for a quantum field appears as the coordinate of a quantum mechanical oscillator. If the kinetic energies of such oscillators, which are proportional to $\dot{a}^2$, are to represent positive energies, we must write our theory in a consistent way, and the replacement $\phi \rightarrow i\phi$ would be wrong.

For electromagnetic waves, it is the components in the transverse direction, perpendicular to the direction of propagation, which are restricted by a similar consideration. A negative sign appears in the associated energy because the energy involves the space indices in the dot product of two vectors, which we have defined as

$$A_\mu B^\mu = A_4 B_4 - (A_3 B_3 + A_2 B_2 + A_1 B_1). \qquad (4.1.6)$$

The sign of the Coulomb forces comes from the sign of the time components in the Lagrangian. For the gravity waves, it is again transverse components that are restricted, but in contracting over two indices (or any even number of indices) the signs cancel out, the sign of the time components h_{44} is the opposite of the electrical case and we have attractions.

4.2   THE STRESS-ENERGY TENSOR FOR SCALAR MATTER

Before we can calculate observable effects and make predictions other than the inverse-square rule and the fact that "likes" attract, with a force proportional to the energy, we must specify how the matter defines the stress tensor $T_{\mu\nu}$. We shall first carry out in some detail calculations based on the simplest assumption, that matter can be represented by a scalar function ϕ. Later, we shall need to consider functions of higher rank; possibly near the end of the course we may consider matter of spin 1/2, because it will have properties significantly different from the integral spins. The integral spins 1 or 2 will require a more complicated algebra, but no conceptual innovations are inherent.

How does one go about generating a stress density from a scalar field ϕ? If we look in Wentzel's book on field theory [Went 49], we find that the following procedure is suggested. The Lagrangian presumably depends on the fields and their derivatives,

$$\mathcal{L} = \mathcal{L}(\psi^i, \psi^i,_\nu). \qquad (4.2.1)$$

The 4−4 component of a stress tensor should represent the energy density, which is the Hamiltonian. Therefore, by a generalization of the usual classical prescription for generating a Hamiltonian from a Lagrangian,

$$H = \dot{q}\frac{\partial L}{\partial \dot{q}} - L, \qquad (4.2.2)$$

the following rule is obtained.

$$T^\mu_{\ \nu} = \psi^i,_\nu \frac{\partial \mathcal{L}}{\partial \psi^i,_\mu} - \delta^\mu_{\ \nu} \mathcal{L} \qquad (4.2.3)$$

This rule is not generally correct. For one thing, it does not necessarily result in an expression symmetric in μ and ν. If the tensor $T_{\mu\nu}$ is not symmetric, the resulting theory is pathological (there is no way to define angular momentum in the field for example). The conservation of energy is confused because the divergences involve terms which are no longer equal,

$$T^{\mu\nu}{}_{,\nu} \neq T^{\nu\mu}{}_{,\nu}. \tag{4.2.4}$$

In our particular scalar case, the rule (4.2.3) does turn out to give a satisfactory symmetric form. We have a Lagrangian and action

$$S(\text{Scalar Matter}) = \frac{1}{2} \int dV (\phi^{,\sigma}\phi_{,\sigma} - m^2\phi^2) \tag{4.2.5}$$

which results in the following stress tensor

$$T_{\mu\nu} = \phi_{,\nu}\phi_{,\mu} - \frac{1}{2}\eta_{\mu\nu}\,\phi^{,\sigma}\phi_{,\sigma} + \frac{1}{2}m^2\phi^2\eta_{\mu\nu}. \tag{4.2.6}$$

With the stress tensor for scalar matter (4.2.6), the coupling term in the Lagrangian becomes

$$-\lambda h^{\mu\nu}T_{\mu\nu} = -\lambda \left[h^{\mu\nu}\phi_{,\mu}\phi_{,\nu} - \frac{1}{2}h^{\mu\nu}\eta_{\mu\nu}(\phi^{,\sigma}\phi_{,\sigma} - m^2\phi^2) \right]. \tag{4.2.7}$$

In terms of our compact notation using bars, this is written as

$$-\lambda \left[\bar{h}^{\mu\nu}\,\phi_{,\mu}\phi_{,\nu} + \frac{1}{2}hm^2\phi^2 \right]. \tag{4.2.8}$$

We may now use such a coupling term to generate amplitudes for scattering by exchange of a graviton.

4.3 AMPLITUDES FOR SCATTERING (SCALAR THEORY)

The amplitude for scattering corresponding to an exchange of a single graviton, diagrammed as in Figure 4.2, may be written by inspection, since we know the form of the propagator, and we have for each vertex the coupling given by the Lagrangian (4.2.7). We replace the gradients by components of the four-momenta in the momentum representation

$$i\phi_{,\nu} = p_\nu \tag{4.3.1}$$

so that the coupling becomes for one vertex

$$2\lambda \left[{}^1p_\mu\,{}^2p_\nu - \frac{1}{2}\eta_{\mu\nu}({}^1p^\sigma\,{}^2p_\sigma - m^2) \right]. \tag{4.3.2}$$

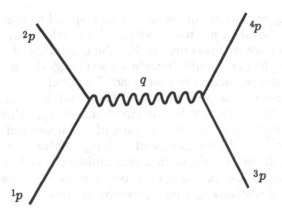

Figure 4.2

We have written underlining under the products $p_\mu p_\nu$ to remind us that we must use an appropriately symmetrized version, since $h_{\mu\nu}$ is symmetric. Explicitly,

$$\underline{A_\mu B_\nu} \equiv \frac{1}{2}[A_\mu B_\nu + A_\nu B_\mu].\tag{4.3.3}$$

For the second vertex we need the "bar" of this expression, which has the form:

$$2\lambda\left[\underline{{}^3p_\mu\,{}^4p_\nu} - \frac{1}{2}m^2\,\eta_{\mu\nu}\right].\tag{4.3.4}$$

The complete expression for the amplitude is then

$$4\lambda^2\left[\underline{{}^3p_\mu\,{}^4p_\nu} - \frac{1}{2}m^2\,\eta_{\mu\nu}\right]\frac{1}{q^2}\left[\underline{{}^1p_\mu\,{}^2p_\nu} - \frac{1}{2}\eta_{\mu\nu}({}^1p^\sigma\,{}^2p_\sigma - m^2)\right].\tag{4.3.5}$$

These abbreviations (bars, underlining, etc.) will result in simplifications of the algebraic manipulations in the more complicated calculations which are to come, so it is worthwhile to use them.

Our theory has given us an expression for the amplitude of gravitational scattering of one particle by another. In order to compute something of observable size, we must go to very large masses, and in order to observe something not given by the Newtonian law, we need velocities close to the speed of light. We can, for example, calculate the deflection of a small mass of very high velocity ($v \approx c$) as it goes by a star such as the sun. Here, we need to justify the replacement of a sum of amplitudes from all the particles in the star by a single amplitude corresponding to a mass M; the replacement is an approximation, but it gives the correct answer to some kind of first-order. The deflection is $(1 + v^2/c^2)$ larger than the prediction of Newtonian theory.

The preceding result cannot be said to correspond to the deflection of light by the sun, because a photon is not a scalar particle, hence cannot be represented by our scalar mass field ϕ. For the scattering of two identical particles, the amplitude should include an exchange term, but for the case of the star, the particles are clearly not identical.

Our theory has not as yet considered the possibility that we might add a piece of zero divergence to our stress tensor $T_{\mu\nu}$; this would correspond to a different distribution in space of the mass and stress. This and related questions will be discussed at length later. Even for scalar matter, as we shall see, we do have a real ambiguity in the description of $T_{\mu\nu}$. This difficulty also arises in electrodynamics when we attempt to write a coupling of photons to charged vector mesons.

4.4 DETAILED PROPERTIES OF PLANE WAVES. COMPTON EFFECT

We can study the properties of the gravitational waves in the absence of matter; by varying the Lagrangian we obtain the equation

$$h_{\mu\nu,\lambda}{}^{,\lambda} - 2\bar{h}_{\mu\sigma,\nu}{}^{,\sigma} = 0 \tag{4.4.1}$$

which is analogous to Maxwell's equation in empty space. If we use plane waves

$$h_{\mu\nu} = e_{\mu\nu} \exp(iq \cdot x) \tag{4.4.2}$$

the equation becomes

$$q^2 e_{\mu\nu} - q_\nu q^\sigma \bar{e}_{\sigma\mu} - q_\mu q^\sigma \bar{e}_{\sigma\nu} = 0. \tag{4.4.3}$$

We are interested in cases when $q^2 \neq 0$ and $q^2 = 0$. If $q^2 \neq 0$ we may divide through by q^2 and rearrange so that

$$e_{\mu\nu} = q_\nu \left(\frac{1}{q^2} q^\sigma \bar{e}_{\sigma\mu} \right) + q_\mu \left(\frac{1}{q^2} q^\upsilon \bar{e}_{\sigma\nu} \right). \tag{4.4.4}$$

This separation has explicitly expressed $e_{\mu\nu}$ as a symmetrized gradient of a vector

$$e_{\mu\nu} = \chi_{\mu,\nu} + \chi_{\nu,\mu}. \tag{4.4.5}$$

We have discussed previously how the gauge invariance of the gravitational field means that the addition of a term of this form makes no difference in the physics. It is therefore always possible to add a piece to $e_{\mu\nu}$ so that $e_{\mu\nu} = 0$. We shall call these waves with $q^2 \neq 0$ "gauge waves"; they have no physical effects and can always be removed by a gauge transformation.

If $q^2 = 0$, then eq.(4.4.3) implies that

$$q^\sigma \bar{e}_{\sigma\mu} = 0. \tag{4.4.6}$$

That is free waves *must* satisfy the Lorentz gauge condition. It is not just a matter of choosing

$$\bar{h}^{\mu\nu}{}_{,\nu} = 0 \tag{4.4.7}$$

for convenience for cases in which the wave is not free. This has its electromagnetic analogue, for photons $q^\mu e_\mu$ must be zero.

We may deduce the actual form of the polarization tensor $e_{\mu\nu}$ in a system of coordinates such that the momentum vector is

$$q^\mu = (\omega, \omega, 0, 0). \tag{4.4.8}$$

If we choose

$$e'_{\mu\nu} = e_{\mu\nu} + q_\mu \chi_\nu + q_\nu \chi_\mu \tag{4.4.9}$$

and demand that $e'_{\mu\nu}$ should have components only in the transverse direction, we obtain a system of equations which may be solved to get the answer

$$e'_{11} = -e'_{22} = \frac{1}{\sqrt{2}} \qquad e'_{12} = e'_{21} = \frac{1}{\sqrt{2}}. \tag{4.4.10}$$

To arrive at eq.(4.4.10), note that the equation (4.4.6) implies that $\bar{e}_{\mu 4} = -\bar{e}_{\mu 3}$ so that only the components 4, 1, and 2 are independent. The 4 components can be removed, if desired, by the transformation (4.4.9). For example, $e'_{14} = e_{14} + \omega \chi_1$ so choose $\chi_1 = -e_{14}/\omega$, $\chi_2 = -e_{24}/\omega$. Then $e'_{43} = e_{43} + \omega \chi_4 - \omega \chi_3$ so choose $\chi_3 - \chi_4 = -e_{34}/\omega$, then $e'_{43} = \bar{e}'_{43} = \bar{e}'_{44} = \bar{e}'_{33} = 0$. We make $e'_{44} = e_{44} + 2\omega \chi_4 = 0$ by choosing $\chi_4 = -e_{44}/2\omega$. Then, since $\bar{e}'_{44}$ is also zero, the trace $e'^\sigma{}_\sigma$ is zero, as is therefore e'_{33} and $e'_{11} + e'_{22}$. Therefore only components with $\mu, \nu = 1$ or 2 in $e'_{\mu\nu}$ survive, and for them $e'_{11} = -e'_{22}$. There are then only two linearly independent normalized combinations (4.4.10).

The amplitude for Compton scattering of a graviton by a particle of mass m corresponds to the diagrams of Figure 4.3. The graviton polarization is represented by the tensor $e_{\mu\nu}$; for scalar mass, the momentum components at each vertex are ${}^1p_\mu$, $({}^1p_\nu + {}^1q_\nu) = ({}^2p_\nu + {}^2q_\nu)$, and ${}^2p_\mu$. In terms of these quantities, we have (for the first diagram)

$$4\lambda^2 \, {}^2\bar{e}^{\mu\nu} \left[{}^2p_\mu \left({}^2p_\nu + {}^2q_\nu \right) - \frac{1}{2} m^2 \eta_{\mu\nu} \right] \frac{1}{({}^1p + {}^1q)^2 - m^2}$$

$$\times \, {}^1\bar{e}^{\alpha\beta} \left[{}^1p_\alpha \left({}^1p_\beta + {}^1q_\beta \right) - \frac{1}{2} m^2 \eta_{\alpha\beta} \right]. \tag{4.4.11}$$

The propagator we have written is that which is appropriate to a scalar particle. Some simplifications in the formula result from the restriction on the plane waves $q^2 = 0$ and $q^\nu \bar{e}_{\nu\mu} = 0$.

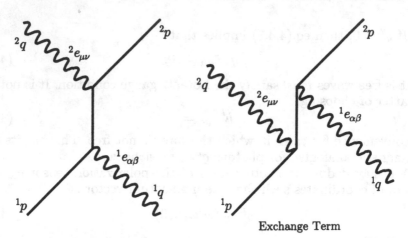

Exchange Term

Figure 4.3

4.5 NONLINEAR DIAGRAMS FOR GRAVITONS

From gauge invariance we expect that a change of $^1e_{\mu\nu}$ to $^1e_{\mu\nu} + {}^1q_\mu a_\nu + {}^1q_\nu a_\mu$ would have no effect on the Compton amplitude. A direct substitution shows that is *not* true. Our result is unsatisfactory and incomplete. What is wrong?

In the Compton scattering of photons by electrons, there is a third diagram, Figure 4.4, not analogous to one of the pair in Figure 4.3. It corresponds to the quadratic coupling, in A^2, which appears in the Lagrangian to make the theory gauge invariant. In analogy to the situation in electrodynamics, we might suspect that in considering only the pair of diagrams, Figure 4.3, we have made an approximation to the truth by linearizing. The existence of an amplitude with quadratic coupling, Figure 4.4, may be deduced in electrodynamics by demanding that the gauge substitution

$$e_1' = e_1 + qa \tag{4.5.1}$$

should lead to no change in the amplitude to a given order. The procedure is simply to equate the terms of the same order of the amplitudes obtained with e_1 and e_1', with coefficients in front of each term which are to be determined. It may be possible to deduce the form of the quadratic graviton terms in an analogous fashion, but this has not as yet been done because the self-coupling of the graviton makes things very complicated in second order, and we shall obtain the correct forms another way around.

It might be interesting to try to deduce these terms in this direct way someday, so we shall make a few remarks on it.

If we consider adding to the Compton scattering, not only amplitudes such as that represented by Figure 4.4, but also the diagram in Figure

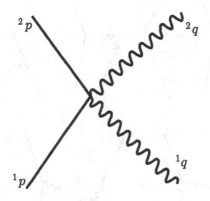

Figure 4.4

4.5, we probably would not have conditions to determine all of the unknown parameters of the more complete theory. If we consider instead the
Compton scattering of a virtual graviton, we may increase the number
of adjustable quantities, and it may be possible to recover the correct
theory. The diagrams involved might be of the type in Figure 4.6 and we
might attempt to make the sum gauge invariant. Actually, we shall solve
these problems in a different way, yet it might be profitable to learn the
details of our field theory by approaching the solutions by different paths.

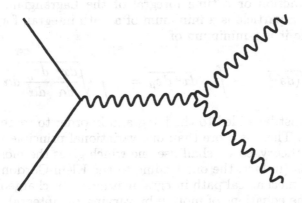

Figure 4.5

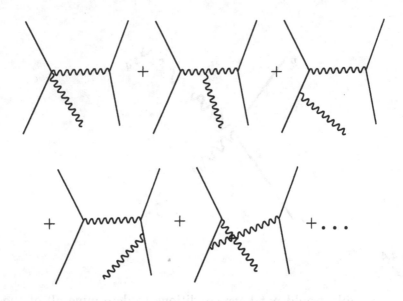

Figure 4.6

4.6   THE CLASSICAL EQUATIONS OF MOTION OF A GRAVITATING PARTICLE

In order to compute some classical effects in our theory, for example, the orbits of planets in going around a star, we need to reduce our quantum theory to its classical form. It is possible to do this by writing the classical theory as the result of a variational principle on a path integral which involves the action or a time integral of the Lagrangian. The motion described by a particle is a minimum of a path integral, for example for a free particle it is a minimum of

$$-\int \sqrt{(ds)^2} = -\int \sqrt{dx^\mu \, dx_\mu} = -\int \sqrt{\frac{dx^\mu}{d\alpha} \frac{dx_\mu}{d\alpha}} \, d\alpha. \qquad (4.6.1)$$

Something must be added to the integrand in order to represent gravitational effects. There is more than one variational principle that can give the classical theory, so we shall use one which gives the most convenient path integrals (in fact, the one leading to the Klein-Gordon equation by the quantum mechanical path integral method). For charged particles we can obtain the equations of motion by varying the integral

$$-\frac{m}{2} \int d\alpha \left(\frac{dx^\mu}{d\alpha}\right) \left(\frac{dx_\mu}{d\alpha}\right) - e \int d\alpha A_\mu(x) \left(\frac{dx^\mu}{d\alpha}\right). \qquad (4.6.2)$$

When we have done some manipulations, the result is

$$m \frac{d^2 x_\mu}{d\alpha^2} = e F_{\mu\nu} \left(\frac{dx^\nu}{d\alpha} \right),$$

(4.6.3)

where $F_{\mu\nu}$ is the curl of A_μ. From this equation, by multiplication by $dx_\mu/d\alpha$, since $F_{\mu\nu}$ is antisymmetric we find that

$$\frac{d}{d\alpha} \left(\frac{dx^\mu}{d\alpha} \frac{dx_\mu}{d\alpha} \right)$$

is zero, or

$$\frac{dx^\mu}{d\alpha} \frac{dx_\mu}{d\alpha} = \left(\frac{ds}{d\alpha} \right)^2$$

is a constant, so α is proportional to proper time (and we can take it equal to proper time if m_0 is the rest mass of the particle). Next we must include our tensor $T_{\mu\nu}$ into the integrand in an appropriate fashion to obtain the correct gravitational equations. In electrodynamics, the vector involved with the field is simply the derivative of a displacement with respect to a four scalar, i.e., a velocity $(dx^\mu/d\alpha)$. We guess that the tensor $T_{\mu\nu}$ is nothing but the tensor generated by two such velocities, and adjust a constant factor so that the 44 component correctly represents the energy density. We set

$$T^{\mu\nu} = m_0 \left(\frac{dx^\mu}{d\alpha} \right) \left(\frac{dx^\nu}{d\alpha} \right),$$

(4.6.4)

where $\alpha = s =$ "proper time." The 44 component is indeed an energy density; it has one factor $1/\sqrt{1 - v^2/c^2}$ to account for the increase of energy with speed, and another to account for the simultaneous shrinkage in volume because of the Lorentz contraction.

The Lagrangian integral or action to be varied is therefore

$$m_0 \left[-\frac{1}{2} \int d\alpha \left(\frac{dx^\mu}{d\alpha} \right) \left(\frac{dx_\mu}{d\alpha} \right) - \lambda \int d\alpha \, h_{\mu\nu}(x) \left(\frac{dx^\mu}{d\alpha} \right) \left(\frac{dx^\nu}{d\alpha} \right) \right].$$

(4.6.5)

We introduce a new tensor to describe this in more compact form,

$$g_{\mu\nu}(x) = \eta_{\mu\nu} + 2\lambda h_{\mu\nu}(x)$$

(4.6.6)

so that our Lagrangian integral becomes

$$m_0 \left[-\frac{1}{2} \int d\alpha \, g_{\mu\nu}(x) \, x'^\mu x'^\nu \right]$$

(4.6.7)

From now on we indicate derivatives with respect to α by primes. As we vary with respect to the path coordinates x, we obtain two equal terms from each of the factors x'^μ, x'^ν, and one from the tensor $g_{\mu\nu}$; the equation of motion is

$$-\frac{d}{d\alpha}\left(g_{\sigma\nu}x'^\nu\right) + \frac{1}{2}\frac{\partial g_{\mu\nu}}{\partial x^\sigma}x'^\mu x'^\nu = 0. \tag{4.6.8}$$

There are other ways of writing the equations that are useful sometimes. First we rearrange terms involving two velocities on one side;

$$g_{\sigma\nu}x''^\nu = \left[\frac{1}{2}\frac{\partial g_{\mu\nu}}{\partial x^\sigma} - \frac{\partial g_{\sigma\nu}}{\partial x^\mu}\right]x'^\mu x'^\nu. \tag{4.6.9}$$

Now we split the second term into two equal pieces and interchange the dummy indices $\mu \leftrightarrow \nu$ in one part, to obtain a combination which is given a special symbol because it is often recurring.

$$[\mu\nu, \sigma] = \frac{1}{2}\left[\frac{\partial g_{\mu\sigma}}{\partial x^\nu} + \frac{\partial g_{\nu\sigma}}{\partial x^\mu} - \frac{\partial g_{\mu\nu}}{\partial x^\sigma}\right] \tag{4.6.10}$$

In terms of this bracket (called "covariant connection coefficients"), the equation of motion is simply

$$g_{\sigma\nu}\, x''^\nu = -[\mu\nu, \sigma]x'^\mu\, x'^\nu. \tag{4.6.11}$$

There is one consequence of the equation which is immediately obtainable by differentiating with respect to α the product $g_{\mu\nu}x'^\mu x'^\nu$.

$$\frac{\partial}{\partial\alpha}\left(g_{\mu\nu}x'^\mu x'^\nu\right) = 2g_{\mu\nu}x'^\mu x''^\nu + \frac{\partial g_{\mu\nu}}{\partial x^\sigma}x'^\mu x'^\nu x'^\sigma. \tag{4.6.12}$$

If we rewrite the product $g_{\mu\nu}x''^\nu$ in the first term by its expression (4.6.9) and relabel indices of summation, we find that the derivative is identically zero. Thus the product $g_{\mu\nu}x'^\mu x'^\nu$ is a scalar constant. If we define a new parameter s by

$$g_{\mu\nu}x'^\mu x'^\nu = \left(\frac{ds}{d\alpha}\right)^2,$$

this s is the analogue for gravity problems of the proper time. Since $ds/d\alpha$ is a constant, we choose it to be unity, and represent all derivatives with respect to s, hereafter, with a dot. In particular, then

$$g_{\mu\nu}\,\dot{x}^\mu\,\dot{x}^\nu = 1. \tag{4.6.13}$$

4.7   ORBITAL MOTION OF A PARTICLE ABOUT A STAR

The equation of motion may be written in terms of the field tensor $h_{\mu\nu}$ as follows (from (4.6.8)):

$$\frac{d}{ds}\left(\eta_{\sigma\nu}\dot{x}^{\sigma} + 2\lambda h_{\sigma\nu}\dot{x}^{\sigma}\right) = \lambda\frac{\partial h_{\mu\sigma}}{\partial x^{\nu}}\dot{x}^{\mu}\dot{x}^{\sigma}. \qquad (4.7.1)$$

Before solving for the motion, we need appropriate expressions for the gravitational fields. We are interested in a region where there are no mass sources. Thus, the field equation

$$h_{\mu\nu,\lambda}{}^{,\lambda} - 2\overline{h}_{\mu\lambda,\nu}{}^{,\lambda} = -\lambda\overline{T}_{\mu\nu}, \qquad (4.7.2)$$

may be solved in a manner analogous to the solution of Maxwell's equations, if we are in the Lorentz gauge $\overline{h}_{\mu\lambda}{}^{,\lambda} = 0$. We have the D'Alembertian $(\square^2 = (\partial/\partial t)^2 - \nabla^2)$

$$\square^2 h_{\mu\nu} = -\lambda\overline{T}_{\mu\nu}. \qquad (4.7.3)$$

For the gravitostatic case, that the time dependence has zero frequency, we must have the Newtonian law of force; the T_{44} component is proportional to the mass. The other components of the stress are zero. The field tensor is

$$\overline{h}_{44} = -\frac{\lambda M}{4\pi r}, \qquad \overline{h}_{\mu\nu} = 0 \quad (\nu,\mu \neq 4,4). \qquad (4.7.4)$$

The unbarred tensor is obtained by barring both sides.

$$h_{\mu\nu} = \overline{h}_{\mu\nu} - \frac{1}{2}\overline{h}^{\sigma}{}_{\sigma}\eta_{\mu\nu} = \begin{cases} -\frac{\lambda}{8\pi}\frac{M}{r} & \text{if } \mu = \nu \\ 0 & \text{otherwise} \end{cases} \qquad (4.7.5)$$

We substitute this field tensor into the equations of motion, and use the following abbreviations

$$\begin{aligned} \phi &= 2\lambda h_{44} \\ \psi &= 2\lambda h_{33} = 2\lambda h_{22} = 2\lambda h_{11}. \end{aligned} \qquad (4.7.6)$$

For the case at hand, $\phi = \psi = -2MG/r$, but we shall have occasion in a later section to solve a case for which this is not true, so we keep the distinction in the algebra, but assume ϕ and ψ are functions of r only.

The procedure for solving for the orbits is analogous to the usual method of solving the equations for a Newtonian field. We separate the equation into space coordinates and time coordinates, eliminate the time and the parameter α to get a differential equation connecting the radius and angular displacement. We start from the four dimensional equation

(4.7.1). The space components ($\nu = 3, 2, 1$) behave according to equations of the form

$$\frac{d}{ds}\left(-\dot{x} + \psi\dot{x}\right) = \frac{1}{2}\left[\frac{\partial\phi}{\partial x}\dot{t}^2 + \frac{\partial\psi}{\partial x}\left(\dot{x}^2 + \dot{y}^2 + \dot{z}^2\right)\right]. \qquad (4.7.7)$$

The time equation is

$$\frac{d}{ds}\left(\dot{t} + \phi\dot{t}\right) = 0. \qquad (4.7.8)$$

We have an integral of the motion coming from the equation (4.6.13),

$$\dot{x}^\mu\dot{x}_\mu + 2\lambda h_{\mu\nu}\dot{x}^\mu\dot{x}^\nu = 1, \qquad (4.7.9)$$

which results in a relation for our case:

$$\dot{t}^2(1 + \phi) - (1 - \psi)(\dot{x}^2 + \dot{y}^2 + \dot{z}^2) = 1. \qquad (4.7.10)$$

The time equation (4.7.8) implies that

$$(1 + \phi)\frac{dt}{ds} = K \qquad (4.7.11)$$

where K is a constant (proportional to the energy). This serves to eliminate the derivative (dt/ds) from the space equation (4.7.7). Since ϕ, ψ depend only on r, the right-hand side of (4.7.7) is directed along x. From this it follows that

$$\frac{d}{ds}\left((1 - \psi)(\dot{x}y - \dot{y}x)\right) = 0.$$

Thus, if we suppose the motion to lie entirely in the plane $z = 0$ and use polar coordinates r, θ in the xy plane we have an additional constant of motion L related to angular momentum:

$$(1 - \psi)\,r^2\dot{\theta} = L \qquad (4.7.12)$$

The radial motion can then be obtained from the equation (4.7.10) in radial coordinates

$$\frac{K^2}{1 + \phi} - (1 - \psi)\left(r^2\dot{\theta}^2 + \dot{r}^2\right) = 1. \qquad (4.7.13)$$

Changing into derivatives $(dr/d\theta)$ by dividing (dr/ds) by $(d\theta/ds)$, we obtain the differential equation for the orbit

$$\frac{K^2}{1 + \phi} - \frac{L^2}{(1 - \psi)r^4}\left[\left(\frac{dr}{d\theta}\right)^2 + r^2\right] = 1. \qquad (4.7.14)$$

The traditional substitution $u = 1/r$ results in an equation that we can conveniently treat by small perturbations on the Newtonian solutions

$$\left(\frac{du}{d\theta}\right)^2 + u^2 = \left(\frac{K^2}{1+\phi} - 1\right)\frac{1-\psi}{L^2}. \qquad (4.7.15)$$

We expect $\phi = \psi = -2MG/r = -2MGu$. From non-relativistic motions, K is near 1 and $K^2/(1+\phi) - 1 = K^2 - 1 + 2MGu$ if ϕ is assumed small, so in that limit of small ϕ, ψ the right-hand side of $(4.7.15)$ is just $L^{-2}(K^2-1+2MGu)$. This is the same as Newtonian theory where the right-hand side is $(E + 2MGu)L^{-2}$ where E is the energy. There are modifications in the relativistic case where we do not neglect terms of higher order. These we discuss in the next section.

5.1 PLANETARY ORBITS AND THE PRECESSION OF MERCURY

As we progress in developing more sophisticated theories, we have to look at progressively finer details of the predictions in order to have criteria to evaluate our theory. We have a field theory that reduces to Newtonian theory in the static limit, which involves the total energy content, and seems to predict correctly the "fall" of photons in the field of a star. The experimental evidence that will make us reject it concerns the precession of the perihelion of the orbit of the planet Mercury. We proceed to compute planetary orbits. We start from the equation

$$
\left(\frac{du}{d\theta}\right)^2 + u^2 = \left(\frac{K^2 - 1 - \phi}{1 + \phi}\right)\left(\frac{1 - \psi}{L^2}\right)
$$

$$
u = \frac{1}{r}; \quad K = (1 + \phi)\frac{dt}{ds}; \quad r\left(\frac{d\theta}{ds}\right) = L = (1 - \psi)r^2\frac{d\theta}{ds},
$$

(5.1.1)

where the symbols ψ and ϕ represent diagonal elements of the tensor $h_{\mu\nu}$, $\phi = 2\lambda h_{44}$ and $\psi = 2\lambda h_{ii}$, $i = 1, 2, 3$. According to our theory, as at present developed, we have $\phi = \psi = -2GM/r = -2GMu$. However, as we shall soon see, our theory is incorrect, so to avoid doing all the work over again when we correct it we shall write

$$\phi = \alpha(-2MGu) + a(-2MGu)^2 + \dots$$
$$\psi = \beta(-2MGu) + b(-2MGu)^2 + \dots \qquad (5.1.2)$$

in our equations, but to find the consequences of our present theory we must put $a = b = 0$, and $\alpha = \beta = 1$ into the formulae at the end. In the case of our scalar theory $\alpha = \beta = 1$, and $\phi = -2MG/r$. We shall assume that the potential ϕ in natural units of the problem, mc^2, is much smaller than 1, so that we may expand the factor $1/(1+\phi)$ as a series in ϕ; the equation becomes

$$\left(\frac{du}{d\theta}\right)^2 + u^2 = \frac{1}{L^2}\left(K^2 - 1 - \phi\right)\left(1 - \phi + \phi^2 \dots\right)\left(1 - \psi\right). \quad (5.1.3)$$

We may now rewrite the right-hand side of this equation as an expansion in terms of the variable u. Keeping only the first and second powers of the small potential, $2GMu$ and $K^2 - 1$, we have

$$\left(\frac{du}{d\theta}\right)^2 + u^2 = A + Bu + Cu^2 + \dots \qquad (5.1.4)$$

where

$$A = \frac{1}{L^2}\left(K^2 - 1\right);$$
$$B = \frac{2GM}{L^2}\left[\left(K^2 - 1\right)\left(\alpha + \beta\right) + \alpha\right];$$
$$C = \frac{1}{L^2}(2MG)^2\left[K^2\alpha^2 + K^2\alpha\beta - K^2a - \left(K^2 - 1\right)b\right].$$

Now we differentiate with respect to θ; after cancelling common factors, the equation has the form which allows simple perturbation solutions,

$$\frac{d^2u}{d\theta^2} + u = \frac{1}{2}B + Cu + \dots. \qquad (5.1.5)$$

When $C = 0$, this equation has the simple conic-section solutions of Newtonian theory. The variable u undergoes harmonic oscillation about the point $1/2\,B$ as a function of θ. For the elliptical orbits, the frequency is 1 so that the radial coordinate r returns to its previous value as the angle θ changes by 2π; the motion is exactly cyclic. When C is not zero, the frequency is $\omega = \sqrt{1 - C}$. The angular period is larger, so that the perihelion returns with an angular change $T = 2\pi/\omega = 2\pi(1 + C/2 + \dots)$.

The angle πC represents the precession of the perihelion per planetary year, since $C \ll 1$.

For a nonrelativistic planet, we obtain the value of the precession easily; the nonrelativistic limit occurs when the total energy K is near 1 (in natural units mc^2). In this case, equation (5.1.5) is easily shown to reduce exactly to the Newtonian equation, as it should. It is more important that our theory should have the correct nonrelativistic limit, than that it give the precession right! When $K^2 - 1 \approx 0$, the precession per planetary year is

$$\pi C = \left(\alpha^2 - a + \alpha\beta\right) 4\pi M^2 G^2 L^{-2}. \qquad (5.1.6)$$

With the present theory, $\alpha = \beta = 1$, $a = 0$, the value obtained is 57 seconds of arc per century (of terrestrial years) for the planet Mercury. For the other planets, the values are much smaller, something like 4 seconds of arc per century in the case of the earth. The astronomical observations yield a value $5270''$ arc/century. However, nearly all of this can be accounted as being due to perturbations because of the presence of the other planets. When the corrections (using pure Newtonian theory) are very accurately made, the discrepancy between observed and calculated precession is 41 ± 2 seconds. Our theory has given a result which is clearly too large, by a factor of something like 4/3.

The results are so close that before scrapping our theory we might review carefully the observational evidence and the calculations to make sure that the discrepancies are real. This has been checked many times and the figure 41 ± 2 remains. We might examine the possibility of physical explanations. If there were a hitherto unobserved planet inside Mercury, or if the sun had a sizeable quadrupole mass deformation, that is, if it were more oblate, precessions could occur. When we actually make estimates as to how large such a deformation would need to be, we arrive at numbers which are much too large to be physically reasonable. The sun is rotating much too slowly to have quadrupole moments of sufficient size. Similar estimates have also made an explanation in terms of inner planets unsatisfactory. It must be concluded that the present theory is not right.

Before proceeding with a discussion of what is wrong with our present theory, we may use the machinery of the orbits to obtain a quantitative result for the deflection of very fast particles in passing near the sun. The relativistic limit is obtained when $K^2 \gg 1$, the total energy is much larger than the rest energy. The linear momentum is $p = \sqrt{K^2 - 1}$. In the limit $K \gg 1$, the equations may be shown to reduce to a form identical to that of Newtonian theory, except that the potential is multiplied by a factor $(\beta + \alpha)$. Since $\beta = \alpha$, the predictions are in general for effects that are twice as large.

The numerical prediction is that very fast particles ($v = c$) grazing the surface of the sun should be deflected by $1.75''$. Measurements have been made on deflection of light from other stars as it passes by the sun, and the results are encouragingly close. The observations are in principle straightforward but it is rather difficult to see *any* stars when the sun is out, let alone close to the stars. Accurate pictures of regions of the sky are taken, to be compared with pictures taken during a total eclipse. When the field of stars is superposed, one tries to match the field of stars far from the sun and detect a shift away from the sun as we get closer in. The analysis of data may be quite lengthy; two such experiments have given results corresponding to deflections of 2.01 and 1.70 seconds of arc for light grazing the sun, so that the prediction 1.75 seconds is generally consistent with the observations.

5.2   TIME DILATION IN A GRAVITATIONAL FIELD

We have at this point a theory which apparently agrees with observations except that the precession of the planetary orbits is overestimated by a factor of 4/3. We may imagine, as Venutian theorists, that while the debate on the residual perturbations goes on or while refined measurements are being made, it is wise to pursue the theory in its present form to discover some new effects that might be tested, or to discover hidden inconsistencies in the theory.

If we compare the differential equations of motion of particles in electrical and gravitational fields we find that the gravitational equation has a qualitatively distinct new feature; not only the gradients, but also the potentials themselves appear in the equations of motion.

$$\text{Elect}: \quad \frac{d^2 x^\mu}{ds^2} = -\frac{e}{m_0} \left(\frac{\partial A^\mu}{\partial x^\nu} - \frac{\partial A_\nu}{\partial x_\mu} \right) \frac{dx^\nu}{ds}$$

$$\text{Gravit}: \quad \frac{d}{ds} \left(g_{\alpha\beta} \frac{dx^\beta}{ds} \right) = \frac{1}{2} \frac{\partial g_{\mu\nu}}{\partial x^\alpha} \frac{dx^\mu}{ds} \frac{dx^\nu}{ds} \qquad (5.2.1)$$

Thus, even though the differential equations for the fields themselves are closely parallel, there is a distinction in the interpretation. For example, the equations do not say the same thing in a region of constant potential and in a region of zero potential, even though the accelerations in either case are zero. In the universe, the contribution to the potential due to faraway nebulae must be very nearly constant over large regions of space, so these considerations apply.

We return to the formulation in terms of the Lagrangian and a variational principle in order to see the new relations with greatest ease and generality. We shall assume that in a certain subregion of space the gravitational tensor $g_{\mu\nu}$ is independent of the coordinates and has the value

$$g_{44} = 1 + \epsilon; \qquad g_{11} = g_{22} = g_{33} = -1. \tag{5.2.2}$$

We expect a negative potential due to faraway masses, $\epsilon < 0$. The Lagrangian is the following:

$$-\frac{m_0}{2} \int d\alpha \, g_{\mu\nu} \frac{dx^\mu}{d\alpha} \frac{dx^\nu}{d\alpha}$$

$$= -\frac{m_0}{2} \int d\alpha \left[(1 + \epsilon) \left(\frac{dt}{d\alpha} \right)^2 - \left(\frac{dx}{d\alpha} \right)^2 - \left(\frac{dy}{d\alpha} \right)^2 - \left(\frac{dz}{d\alpha} \right)^2 \right] \tag{5.2.3}$$

It is evident that the simple substitution $t' = t\sqrt{1 + \epsilon}$ restores this term to its previous algebraic form. This would seem to say that the effect of a constant potential is like a change in the scale of time to make physical processes run more slowly in regions of lower gravitational potential.

The argument in terms of the free particle only is not meaningful, since we cannot very well claim that the rate at which nothing is happening can change. We must look at the behavior of interacting particles. For this purpose, we continue using our theory of scalar matter; the action integral is

$$\frac{1}{2} \int d^4x \left(\phi^{,\sigma}\phi_{,\sigma} - m^2\phi^2 \right) - \lambda \int d^4x \, h_{\mu\nu}T^{\mu\nu} \tag{5.2.4}$$

where

$$T^{\mu\nu} = \phi^{,\mu} \frac{\partial \mathcal{L}}{\partial \phi_{,\nu}} - \eta^{\mu\nu} \mathcal{L}. \tag{5.2.4'}$$

We can explicitly separate the space derivatives and time derivatives in the gradients, and also separate the time in the volume element d^4x. We assume that the corrections ϵ are smaller than 1 so that an expansion is permissible, and arrive at the following expression for the action integral

$$\frac{1}{2} \int d^3x \, dt \left[\left(\frac{\partial \phi}{\partial t} \right)^2 (1 - \epsilon/2) - (\nabla\phi)^2 (1 + \epsilon/2) - m^2\phi^2 (1 + \epsilon/2) \right]. \tag{5.2.5}$$

Again it happens that for $dt' = dt\sqrt{1 + \epsilon} \approx dt(1 + \epsilon/2)$ the action is restored to its previous algebraic form. Clearly, the time dilation occurs for our scalar mesons represented by ϕ. It is possible to show that the time dilation should occur for all interactions, regardless of the exact nature of the total Lagrangian. We can argue with the help of Wenzel's formula, (5.2.4'), for $T^{\mu\nu}$. The gravitational interaction may be explicitly separated from the rest of the Lagrangian, whatever it is.

$$\mathcal{L}(\text{total}) = \mathcal{L}_0 - \lambda h_{\mu\nu} T^{\mu\nu} \qquad (5.2.6)$$

Using formula (5.2.4') and $g_{\mu\nu}$ as before (5.2.2) so $\lambda h_{44} = \epsilon/2$, the total Lagrangian is $\mathcal{L} - (\epsilon/2)T_{44}$ or

$$\mathcal{L}(\text{total}) = \mathcal{L}(1 + \epsilon/2) - \frac{\partial \mathcal{L}}{\psi_{,t}}\,\psi_{,t}(\epsilon/2). \qquad (5.2.7)$$

Let us assume therefore that the total Lagrangian (including our constant gravitational potential) involves only the field ψ and its gradients. The action integral in terms of the variable t' is, at least to first order in ϵ,

$$\text{Action} = \int d^3x\, dt'\, \mathcal{L}\left(\psi_{,t'}, \psi, \psi_{,x}\right) \qquad (5.2.8)$$

since

$$\psi_{,t'} = (1 + \epsilon)^{-1/2}\psi_{,t}$$

and

$$\text{Action} = \int d^3x\,(1 + \epsilon)^{1/2}dt\, \mathcal{L}\left[(1 + \epsilon)^{-1/2}\psi_{,t}, \psi, \psi_{,x}\right].$$

The result of all this is that any terms in the Lagrangian involving the time gradients $\psi_{,t}$ carry their own factors $\sqrt{1 + \epsilon}$ so that the substitution $t' = t\sqrt{1 + \epsilon}$ exactly reproduces the effect of the constant gravitational field. All physics therefore remains the same except for the time dilation.

The gravitational potentials are negative, so that clocks should run more slowly as they come nearer to a massive object such as a star. One might ask whether there is a possibility that $(1 + \epsilon)$ should be negative, since $\epsilon = -2GM/r$. In practice, the question never arises because G is so small. For a star of the mass of the sun, we would have $\epsilon = -1$ only if the mass were concentrated inside a sphere of some 1.5 kilometers radius. However the mathematical possibility of $\epsilon < -1$ exists in our theory, and we shall discuss later how even a refined theory produces such difficulties.

Thus, here we have a new prediction of our gravitational theories; clocks should run more slowly in regions of lower gravitational potential. The terrestrians have carried out this experiment, by dropping photons near the surface of the earth a distance of 24 meters. The photons are emitted at the top and absorbed at the bottom; one uses the extremely sharp lines discovered by Mössbauer coming from nuclear transitions in crystals. The small change in frequency due to the drop (1 part in 10^{15}) is compensated by an artificial Doppler shift. When the absorption as a function of the relative velocities of the crystals is used to determine the frequency shift, the results agree with the theoretical predictions within the experimental uncertainty of some ten percent. The clocks which run more slowly in this case are nuclear mechanisms which produce photons of definite frequencies; the fractional difference in the frequencies of the clocks at the top and bottom is $(\Delta\omega/\omega) = (\epsilon/2) =$ the difference of the gravitational potential divided by c^2.

The prediction of this frequency shift does not really need the machinery of our theory of gravitation, since it is implicit in the experimental results of Eötvös, that gravity forces (or potentials) are proportional to the energy content. Thus, the frequency shift corresponds to the gravitational energy of the photon energy. According to Eötvös, the excited nucleus is heavier by $(E_0/c^2)g$, if E_0 is the excitation energy, since as we know from nuclear experiments its mass is $M + E_0/c^2$, if M is the mass in the ground state. When it is raised by a height h it contains an energy $E_0 + (E_0/c^2)gh + Mgh$ more than an unexcited nucleus at zero height. If we excite the lower nucleus, we require only E_0. After the upper nucleus makes a transition, its total energy should exceed that of the lower nucleus only by Mgh. Since the photon frequency is $E = \hbar\omega$, the frequency of the photon emitted is $\omega = \omega_0[1 + gh/c^2]$. It is thus obvious that the frequency shift is required by energy conservation. If this shift were not there, we might yet construct a perpetual-motion machine, using such nuclear transitions. We excite nuclei at the top by photons of energy E_0, but we get a mechanical energy $(M + E_0/c^2)gh$ as we lower an excited nucleus. Since taking up unexcited nuclei costs us only Mgh, we get $(E_0/c^2)gh$ free with each cycle! Our theory is then not inconsistent, and it suggests that the frequency shift required by energy conservation be considered as a general property of all physical processes, that they run more slowly in lower gravitational potentials.

There is nothing like the "twin paradox" of special relativity here. The man on top of a mountain is living and aging at a faster rate than we; we see him move faster. When he looks back at us, he sees us moving more slowly than he. It is not like the time dilation of high relative velocities, when each observer sees the other moving slowly. There is no way to significantly increase our life span by moving to Death Valley, however; the rates of aging change very little. Still, we might have to be more careful in the future in speaking of the ages of objects such as the earth, since the center of the earth should be a day or two younger than the surface!

5.3   COSMOLOGICAL EFFECTS OF THE TIME DILATION. MACH'S PRINCIPLE

Previously, we had noticed that the universe might be roughly described as a spherically symmetric mass distribution, and that the gravitational potentials were possibly of such size that the gravitational energies were equal to the rest energies of particles near the center. If this were so, and if our formula for time dilation were correct, physical processes should stop at the center of the universe, since the time would not run at all. This is not the only physically unacceptable prediction; since we might expect that matter near the edge of the universe should be interacting faster, light from distant galaxies should be violet-shifted. Instead, it is well known to be shifted toward lower, redder frequencies. Thus, our formula for the time dilation obviously needs to be discussed further in connection with possible models of the universe. The following discussion is purely qualitative and is meant only to stimulate wiser thoughts on this subject.

It is conceivable that corrections to our simple formula might come from the space elements of the tensor $h_{\mu\nu}$. We have considered h_{44} only; it may be that if we included h_{33} and h_{22} and h_{11}, we might predict not simply a time dilation but some simultaneous contraction along the space axes that would somehow resolve the difficulties. Another possibility is that the 1 which appears in the formula of time dilation is a mistake in thought. We have written a formula which applies only when the potential differences ϕ are much smaller than 1, so that the constant 1 may somehow represent a normalized contribution of faraway nebulae. In other words, we have deduced that gravitational corrections to the total energy of a particle are corrections to its inertia. It is a conceptually simple extension of the thought to assume that perhaps particles have no intrinsic inertia, so that all inertia represents a sum of gravitational interactions with the rest of the universe. We run into quantitative difficulties immediately. Suppose that we attempt to say that near the sun a single planet has a total potential energy which is the sum of the sun's potential and a nearly-constant contribution due to the rest of matter:

$$\phi = \phi(\text{sun}) + \phi(\text{matter}). \tag{5.3.1}$$

We have no possibility of identifying ϕ(nebulae) with the 1, since the correction ϕ(sun) must be of a different sign.

Even though this unsophisticated attempt appears to fail, it may be worthwhile to discuss this point in more detail. The idea that inertia represents the effects of interactions with faraway matter was first developed by Ernst Mach in the nineteenth century, and it was one of the powerful ideas that Einstein had in mind as he constructed his theory of gravitation.

Mach felt that the concept of an absolute acceleration relative to "space" was not meaningful; that instead, the usual absolute accelerations of classical physics should be rephrased as accelerations relative to the distant nebulae. Similarly, the notion of rotation should be rotation relative to something else, an "absolute rotation" being also a concept devoid of significance. When we consider this statement as a fundamental assumption or postulate, it is known as Mach's Principle. It is possible that all by itself it may lead to meaningful physical results, much in the same way that the principle of relativity, connecting systems of reference in uniform relative velocity, served Huygens as a tool to deduce the laws of collision of billiard balls; suppose we observe in a head-on collision that billiard balls having equal and opposite momenta reverse their momenta. Huygens then imagined the same experiment being performed on a boat, having a uniform velocity relative to the shore. By using the principle of relativity, he arrived at the correct law for the collision of smooth billiard balls having any arbitrary initial velocities.

Mach's Principle would profoundly alter the laws of mechanics, since the usual mechanics assumes unaccelerated rectilinear motion to be the "natural" motion in the absence of forces. When accelerations are defined as accelerations relative to other objects, the path of a particle under "no acceleration" depends on the distribution of the other objects in space, and the definitions of forces between objects would be altered as we change the distributions of other objects in space.

5.4   MACH'S PRINCIPLE IN QUANTUM MECHANICS

The statement of Mach's Principle for quantum theories involves new effects, since we cannot speak of rectilinear paths; we shall see that the proper statement involves rather the development of time.

Mach had the problem of deciding how a particle "knew" it was accelerating. He believed it was due to an influence from the nebulae, an influence such that acceleration relative to them required force. With the advent of quantum mechanics a new "absolute" was definable; absolute scale of length or time. 10^{24} hydrogen atoms at zero energy in a cube has just some definite absolute size, the NH_3 molecule rings with a definite time between cycles. In a vacuum two equal photons heading toward each other do nothing special until the wavelength is less than $2\pi\ 3.68 \times 10^{-11}$ cm when pairs of electrons can be produced. How do the photons "know" what their wavelength is in absolute units to decide whether to make pairs? Each volume of space must contain a natural measure of size (or time).

Accepting the philosophy of Mach's Principle, we would say that the above is nonsense, the size is not absolute if there is nothing to compare it to. It must be the influence of the nebulae which determines the scale of time at each point in space. Say the Compton wavelength relative to the size of the universe depends on how many nebulae are in it. If they were partly removed, the scale would change, presumably.

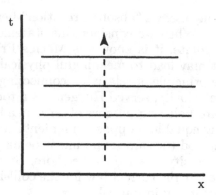

Figure 5.1

We therefore assume that the natural scale of time—the size say of $\hbar/mc^2$ (or any other combination for other fundamental particles—we suppose they all are proportional to some unit of length)—is determined by the distant nebulae. We shall now show that the inertial frame is now also automatically determined from the nebulae, and the phenomena of inertia for accelerations relative to the nebulae can be understood if the "length determining principle" is accepted. Therefore, Mach's Principle is equivalent to the statement that the fundamental units of length and time at a point are the result of the influence of nebulae.

A particle hypothetically "at rest" in quantum mechanics has a time development $\exp(-imc^2t/\hbar)$. The principle of inertia is a statement that the time scale is independent of coordinates x; the classical trajectories are interpreted to follow the normal lines of constant phase. In two dimensions, we diagram lines of constant phase perpendicular to the time axis, as in Figure 5.1.

If the time scales in different parts of space are not the same, the lines of constant phase in such a diagram are curves, and the corresponding classical paths are curved, corresponding to accelerated motion toward the region of smaller scale as illustrated in Figure 5.2. Since stars produce such a closer packing time, they must produce accelerations. In quantum mechanics, plane-wave solutions exist when the surfaces of constant phase are parallel; if not, the wave packets will tend to follow the gradient of the phase. Now if the distant nebulae mainly determine the scale, and there are no nearby stars, the scale $\hbar/mc^2$ would be almost exactly equal at two nearby points, 1 and 2, because 1 and 2 are almost exactly at the same distance from all the nebulae. Therefore, the natural frequency (separation of lines of constant phase) at 1 and 2 would be practically equal. Thus if a particle had initially equal phase at 1 and 2 it would

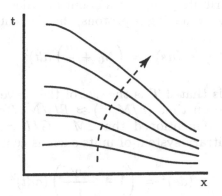

Figure 5.2

always do so, and it would stand still—not accelerate (more correctly, a long wave packet would stand still). Unequal initial phase gives sloping lines, but of a slope constant in time—a uniform velocity. The absence of acceleration is a consequence of the natural time scale being equal at all points in a region of space. This constancy is understood if the nebulae determine the natural scale, for in a space very small compared to the dimensions of the distribution of influencing nebulae (size of the universe) no variation in scale could be expected.

There are some numerical coincidences that we might mention at this point, in order to suggest how the "natural" dimensions of length might arise out of cosmology in some sense. The coincidence involves no "theory" as such—it merely serves to illustrate the type of connection that might eventually be predicted by a detailed theory. We suppose that in some system of time units which is "natural" to the universe, the invariant arc length for a particle at rest is

$$(ds)^2 = g_{44}(dt)^2. \qquad (5.4.1)$$

The coordinate time unit of t might be (R/c), where R is the radius of the universe. We suppose atomic units are defined by ds—*we take the absolute size of $g_{\mu\nu}$ seriously*—one unit of ds is the fundamental length. Which fundamental length? They are all proportional but we try the proton Compton wavelength $\hbar/Mc$. That is, an s of 1 means one oscillation of a proton wave function, a t of 1 is the scale of the size of the universe.

We next assume that the contribution to g_{44} due to each proton is simply $1/r$. (r is in coordinate units, the radius of the universe.) Then the faraway nebulae, which have N_0 protons and which are a typical distance

$R = 1$ away, contribute to g_{44} an amount of order N_0. In the vicinity of a star at a distance r having n protons, the element of arc has a square

$$(ds)^2 = \left(N_0 + \frac{n}{r}\right)(dt)^2. \qquad (5.4.2)$$

The coincidence is that if T is the age of the universe, it is numerically related to the proton time $(\hbar/M_p c^2) \approx T/\sqrt{N_0}$. Together with another coincidence we have mentioned, that $2M_{\text{univ}}G/R \approx 1$, the result when we change to an arbitrary system of units such as centimeters and seconds, is

$$(ds)^2 = \left(1 + \frac{2mG}{r}\right)(c\,dt)^2, \qquad (5.4.3)$$

where m is the mass of the star, $m = nM_p$ roughly. Except for the disastrous appearance of a $(+)$ sign instead of a $(-)$ sign, the result is identical to the "correct" arc length. We have succeeded in getting the correct sizes by juggling purely cosmological numbers.

There is probably no deep significance in this coincidence. One thing that is wrong is that we have assumed a contribution $(1/r)$ from each proton mass, but the $(1/r)$ dependence is the correct form for the corrections to the total energy due to nearby particles, and possibly incorrect for the particles in distant galaxies. Another grave difficulty is that we have made no attempt to include effects of other terms of the tensor $h_{\mu\nu}$, for example h_{11}. Still, the juggling serves to indicate how theories of gravitation inevitably lead to considerations involving times and inertia; we get a glimpse of how an interaction in terms of numbers of distant particles may lead to the observed inertia of an object such as a proton. At any rate, a suggestion is made that the absolute size of $g_{\mu\nu}$ be taken seriously; it may have a meaning. Flat space may be $g_{44} = -g_{11} = -g_{22} = -g_{33} = \xi$, where ξ is a meaningful number, not to be simply taken as 1.

5.5   THE SELF ENERGY OF THE GRAVITATIONAL FIELD

We return to less speculative and more precise matters.

In developing and making modifications in our field theory, we have neglected to check that the theory be internally consistent. We have written down a total Lagrangian having a field term, a matter term, and a coupling term. We have arrived at a field equation by arranging that the divergence of the stress-energy tensor should be identically zero. This procedure is evidently incorrect, since we have written a stress tensor which did not include the energy of the gravitational field itself. Thus, our present theory is physically untenable, since the energy of the matter is not conserved.

We shall attempt to correct this theoretical deficiency by searching for a new tensor to be added to our old $T^{\mu\nu}$, which might fix things up so that

$$(T^{\mu\nu} + \chi^{\mu\nu})_{,\nu} = 0, \qquad (5.5.1)$$

and at the same time the self energy of the field is correctly taken into account. How do we find this term? We might attempt to construct the correct total tensor using Wenzel's formula and the complete Lagrangian. The result is an unsymmetric tensor, if we symmetrize and compute, the precession of the perihelion of Mercury comes out wrong. This is another example of a rule of thumb about theories of physics: Theories not coming from some kind of variational principle, such as Least-Action, may be expected to eventually lead to trouble and inconsistencies.

We shall make a different kind of attempt, in line with our usual procedure of trying out various theories in sequential order of increasing complexity. Physically, we know we are attempting to describe a nonlinear effect: the gravitational field is produced by energies, and the energy of the field is a source of more fields. Here we may start to get excited—it is certainly conceivable that such nonlinearity may account for the small residual discrepancy in the precession of the perihelion of Mercury. We shall insist that the field equations come out of the variation of some action, and shall ask ourselves what kind of a term must we add to the Lagrangian to get a term like $\chi^{\mu\nu}$ to go into an equation of motion

$$h^{\mu\nu,\sigma}{}_{,\sigma} - 2\bar{h}^{\mu}{}_{\sigma}{}^{,\nu\sigma} = -\lambda \left(\overline{T}^{\mu\nu} + \overline{\chi}^{\mu\nu} \right), \qquad (5.5.2)$$

and such that eq.(5.5.1) is satisfied? What might $\chi^{\mu\nu}$ look like if it represents a kind of gravitational energy? Undoubtedly, at least in part, it is proportional to squares of field strengths; that is, the product of two gradients of the potentials. Perhaps, therefore, $\chi^{\mu\nu}$ is a sum of terms like $h^{\mu\sigma}{}_{,\lambda}h^{\nu\lambda}{}_{,\sigma} +$ etc. each with two h's and two derivatives.

We shall insist that our equations be deducible from a variational principle such as Least Action. When we vary the products, we reduce the number of h's, so the Lagrangian to be varied needs a net term, of third order in $h_{\mu\nu}$, which we shall call F^3; we shall try to arrange things so that the variation of F^3 leads to the term $\chi^{\mu\nu}$.

$$\frac{\delta F^3}{\delta h_{\mu\nu}} = \lambda \chi^{\mu\nu} \qquad (5.5.3)$$

The algebraic character of F^3 must be that it involves products of three h's and has two derivative indices. A typical term of F^3 might be

$$F^3 = ah_{\mu\nu}h^{\mu\sigma,\lambda}h_{\sigma\lambda}{}^{,\nu} + \ldots \qquad (5.5.4)$$

When we write all possible such products, we find there are 24. We may further reduce the number by noting that certain terms can be reduced to combinations of others by integrating twice by parts; these considerations lead us to write down 18 different and independent forms. We wind up therefore with an expression for $\chi^{\mu\nu}$ in terms of h's and 18 independent constants.

The next procedure is evident. We attempt to determine the constants by demanding that

$$(T^{\mu\nu} + \chi^{\mu\nu})_{,\nu} = 0. \tag{5.5.5}$$

This results in a set of many more than 18 equations for the 18 constants. However, it turns out that they are all consistent, and the 18 constants are unique. When we have done this, we shall have an improved theory which correctly takes into account the energy of the gravitational field itself to second order in $h_{\mu\nu}$.

6.1 THE BILINEAR TERMS OF THE STRESS-ENERGY TENSOR

Our present theory is linear in the sense that we have written down an equation relating the gravitational fields $h_{\mu\nu}$ to a stress tensor $T_{\mu\nu}$.

$$\overline{h_{\mu\nu,\lambda}{}^{,\lambda} - 2\bar{h}_{\mu\lambda,\nu}{}^{,\lambda}} = -\lambda T_{\mu\nu} \qquad (6.1.1)$$

But we have specified $T_{\mu\nu}$ in terms of matter alone, as though it were unaffected by gravity, and as though the gravitational field energy were not itself a source of fields. The effects of gravity on matter that we want to include may be illustrated by considering what may happen as we bring two masses 1 and 2 together, in the presence of a third object. Part of the work done may go into heating up the third object, so that energy is not conserved by considering only the masses 1 and 2 and the fields they generate. Thus, energy would not be conserved if we consider subsystems only; the dashed boxes shown in Figure 6.1 would not weigh the same. The nonlinear effect due to field energies is conceptually more familiar; we have calculated the fields due to mass sources as a first approximation;

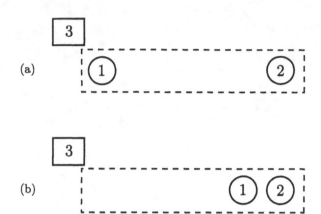

Figure 6.1

the next approximation is to include the first-order fields as sources, and so approach a self-consistent solution.

We shall construct a new stress tensor from our old one by adding a term which shall be derivable from a neglected piece of the Lagrangian, which we denote by F^3, by a variation

$$^{\text{new}}T^{\mu\nu} = {}^{\text{old}}T^{\mu\nu} + \chi^{\mu\nu}$$
$$\lambda\chi^{\mu\nu} = \frac{\delta F^3[h]}{\delta h_{\mu\nu}}, \tag{6.1.2}$$

and hope that the difficulties will be removed, at least to a higher order in $h_{\mu\nu}$.

Since we are attempting to construct $\chi^{\mu\nu}$ so that it fixes up the failure of $^{\text{old}}T^{\mu\nu} \equiv {}^{0}T^{\mu\nu}$ to conserve energy, $^{0}T^{\mu\nu}{}_{,\nu} \neq 0$, we get a hint as to the structure of $\chi^{\mu\nu}$ by taking the divergence $^{0}T^{\mu\nu}{}_{,\nu}$. The divergence of $\chi^{\mu\nu}$ should cancel it out, at least to the first nonvanishing order. To compute this divergence, we first rewrite $^{0}T^{\mu\nu}$ for a moving particle in a new form which looks at first unfamiliar but allows easier manipulations. In terms of an integral over a scalar parameter which might as well be the proper time s (we indicate derivatives with respect to s by dots),

$$^{0}T^{\mu\nu}(x) = m_0 \int ds \, \delta^4(x - z(s)) \, \dot{z}^\mu \dot{z}^\nu. \tag{6.1.3}$$

That this expression for $^{0}T^{\mu\nu}$ is equivalent to what we have used previously may be checked by comparing the corresponding terms of the action

$$\lambda \int d^4x \, ^{0}T^{\mu\nu}(x) h_{\mu\nu}(x) = \lambda m_0 \int ds \, h_{\mu\nu}(z) \, \dot{z}^\mu \dot{z}^\nu. \tag{6.1.4}$$

There is a simple physical way to interpret the significance of the δ-function in eq.(6.1.3); it simply says that there is no interaction energy except where the particle actually is. It is perhaps easier to understand how appropriate these forms are by rewriting our familiar electrodynamics in the same language; the interaction term of the Lagrangian is a volume integral of $-j^\mu A_\mu$ and j^μ is related to the particle velocity

$$j^\mu(x) = e \int ds\, \delta^4(x - z(s))\, \dot{z}^\mu$$

$$S(\text{int}) = -e \int ds\, A_\mu(z)\, \dot{z}^\mu. \qquad (6.1.5)$$

The parallelism with our gravitation-field expressions, eqs.(6.1.3) and (6.1.4), is evident.

We compute the divergence $^0T^{\mu\nu}{}_{,\nu}$ from eq.(6.1.3). We first recognize that the δ-function is symmetric in x and z, so that the derivatives with respect to x^ν in the divergence may be replaced by (minus) derivatives with respect to z^ν. Then we make use of the identity

$$\dot{z}^\nu \frac{\partial}{\partial z^\nu} f[z(s)] = \frac{d}{ds} f[z(s)] \qquad (6.1.6)$$

to obtain for the divergence of $^0T^{\mu\nu}$,

$$^0T^{\mu\nu}{}_{,\nu} = \int ds\, \delta^4(x - z(s))\ddot{z}^\mu. \qquad (6.1.7)$$

We see that this divergence is a density of acceleration. At this point we shall assume that we have already included correctly in our Lagrangian all energies other than gravitation, so that the acceleration $\ddot{z}^\mu$ represents the effect of gravity as given by the equation of motion

$$g_{\mu\lambda}\ddot{z}^\mu = -\frac{1}{2}\left[g_{\mu\lambda,\nu} + g_{\nu\lambda,\mu} - g_{\mu\nu,\lambda}\right]\dot{z}^\mu \dot{z}^\nu$$

$$= -[\mu\nu, \lambda]_z\, \dot{z}^\mu \dot{z}^\nu. \qquad (6.1.8)$$

The subscript z on the bracket is to remind us to which variables the indices refer. We now multiply the divergence, eq.(6.1.7), by $g_{\mu\lambda}(x)$, and replace $g_{\mu\lambda}\ddot{z}^\mu$ by $-[\mu\nu, \lambda]_z\dot{z}^\mu\dot{z}^\nu$. We note that because of the δ-function, $[\mu\nu, \lambda]_z$ has the same effect as $[\mu\nu, \lambda]_x$. This means that the bracket symbol can be pulled out of the integral sign, leaving us an expression involving the divergence $^0T^{\mu\nu}{}_{,\nu}$ and the original tensor $^0T^{\mu\nu}$.

$$g_{\sigma\lambda}(x)\, ^0T^{\sigma\nu}{}_{,\nu}(x) = -[\mu\nu, \lambda]\, ^0T^{\mu\nu}(x) \qquad (6.1.9)$$

This is an exact equation which $^{0}T^{\mu\nu}$ must satisfy. At present we use it
only to first order in h. We may separate $g_{\sigma\lambda}$ into its parts $\eta_{\sigma\lambda} + 2\lambda h_{\sigma\lambda}$,
and obtain an equation which tells us that $^{0}T^{\lambda\nu}{}_{,\nu}$ starts with a linear
term in the coupling constant λ,

$$^{0}T_{\lambda\nu}{}^{,\nu} = -[\mu\nu, \lambda]^{0}T^{\mu\nu} - 2\lambda h_{\sigma\lambda}{}^{0}T^{\sigma\nu}{}_{,\nu}, \qquad (6.1.10)$$

since the bracket symbol involves derivatives which make the zero-order
part $\eta_{\mu\nu}$ of $g_{\mu\nu}$ play no role.

When we compare this equation to the requirement that the new
tensor $^{\text{new}}T^{\mu\nu} \equiv {}^{n}T^{\mu\nu}$ should have zero divergence,

$$^{n}T^{\mu\nu}{}_{,\nu} = {}^{0}T^{\mu\nu}{}_{,\nu} + \chi^{\mu\nu}{}_{,\nu}, \qquad (6.1.11)$$

and if we assume that $\chi^{\mu\nu}$ itself is bilinear in the fields, we see that the
divergence $\chi^{\mu\nu}{}_{,\nu}$ should be

$$\chi_{\mu\nu}{}^{,\nu} = [\sigma\nu, \mu]^{0}T^{\sigma\nu} + \mathcal{O}(\lambda^2) \dots . \qquad (6.1.12)$$

Knowing the divergence does not determine $\chi^{\mu\nu}$ for us. We have an
additional requirement in that we expect to deduce $\chi^{\mu\nu}$ from a variation
of F^3 with respect to $h_{\mu\nu}$, eq.(6.1.2). If we construct F^3 as a sum over all
possible independent products involving trilinear products of field components and two derivative indices, these two requirements determine F^3
uniquely. We shall not carry out here the determination of the 18 constants, but quote the result of a lot of hard algebraic labor.

$$F^3 = -\lambda \left[h^{\alpha\beta} \overline{h}^{\gamma\delta} h_{\alpha\beta,\gamma\delta} + h_\gamma{}^\beta h^{\gamma\alpha} \overline{h}_{\alpha\beta,\delta}{}^{,\delta} \right.$$

$$-2h^{\alpha\beta} h_{\beta\delta} \overline{h}_{\alpha\gamma}{}^{,\gamma\delta} + 2\overline{h}_{\alpha\beta} \overline{h}^{\sigma\alpha}{}_{,\sigma} \overline{h}^{\tau\beta}{}_{,\tau}$$

$$\left. + \left(\frac{1}{2} h_{\alpha\beta} h^{\alpha\beta} + \frac{1}{4} h^\alpha{}_\alpha h^\beta{}_\beta \right) \overline{h}^{\sigma\tau}{}_{,\sigma\tau} \right]. \qquad (6.1.13)$$

It is now possible for us to compute, using perturbation techniques,
all of the effects we have previously considered. For the case of planetary motion, the inclusion of F^3 in the Lagrangian integral produces the
following values of ϕ and ψ to be used in calculations of orbits:

$$\phi = \Phi + \frac{1}{2}\Phi^2 \dots$$

$$\psi = \Phi - \frac{3}{8}\Phi^2 \dots \qquad (6.1.14)$$

$$\Phi = -2MG/r.$$

These corrections result in a perfect agreement of our theory with the
observations on the precession of the perihelion of Mercury, so that the
last remaining discrepancy between theory and observation is removed.

6.2   FORMULATION OF A THEORY CORRECT TO ALL ORDERS

We have succeeded in the task we had set before us in the beginning, to develop a field theory of gravitation in analogy to our other well-known field theories, which would adequately describe all the known characteristics of gravitational phenomena. Thus, our fictional Venutian viewpoint has been fruitful. There are yet some loose ends in our theory; we might conceive that the hard-working among the Venutian theorists might be dissatisfied with a theory which left third-order effects unspecified, and some of them might pursue the chase after functions F^4 and then F^5 to be added to the Lagrangian integral to make the theory consistent to higher orders. This approach is an incredibly laborious procedure to calculate unobservable corrections, so we shall not emulate our fictional Venutians in this respect.

It has happened in physical theories that although high order corrections in a particular expansion are exceedingly tedious to compute, it is possible to construct a theory which sums all higher order corrections to give an answer which is accessible. Thus we imagine an ambitious and bold Venutian who decides to make the attempt to deduce the entire expression for a function $F = F^2 + F^3 + F^4 + F^5 + \ldots$. We shall search for a functional F which is to be an action to be varied, for empirical reasons: There is apparently no successful theory which is not derivable from a variational principle which starts from a Lagrangian or a Hamiltonian functional (which are equivalent).

It is not certain whether the failures of non-Lagrangian theories reflect some fundamental truth about nature. It is possible that the fundamental truth may be that processes occur according to a principle of minimum phase, and that the actions of classical physics or of quantum physics are expressions for this phase which are correct to some approximation. An ambitious attempt to cast gravitation into a non-Lagrangian formulation was made by Birkhoff [Birk 43]. He preserved linear equations for the fields, but changed the equations of motion for particles. The resulting classical theory was quite satisfactory, but it did not allow a consistent quantization. It was shown that the wave motion of wave packets did not follow the postulated classical equations, but rather Einstein's equations! It seems probable that the attempt at quantization uncovered some hidden inconsistency in the field equations.

We shall therefore search for the complete functional F,

$$F = F^2 + F^3 + F^4 + \cdots, \tag{6.2.1}$$

which shall be defined by the requirement that the resulting equation of motion,

$$\frac{\delta F}{\delta h_{\mu\nu}} = \lambda T^{\mu\nu}, \tag{6.2.2}$$

shall automatically imply the divergence property, eq.(6.1.9), of $T^{\mu\nu}$. The functional F must therefore satisfy the following differential functional equation,

$$g_{\sigma\lambda}\left(\frac{\delta F}{\delta h_{\sigma\nu}}\right)_{,\nu} + [\mu\nu,\lambda]\left(\frac{\delta F}{\delta h_{\mu\nu}}\right) = 0, \qquad (6.2.3)$$

which we are to solve. This is in general an exceedingly difficult problem, and there is no procedure for generating the solutions. We shall have to rely on our ingenuity in devising functionals which are solutions in the sense that they satisfy eq.(6.2.2) when plugged in. There is no general unique solution, even if we add that for small h we shall choose that solution whose leading terms are the F^2 and F^3 we have deduced by other methods. There is, however, an evident "simplest" solution (involving the smallest number of derivatives of the $g_{\mu\nu}$—just two). We choose it. When this is done, we shall have arrived at a theory which is identical to Einstein's. At that point, then, we shall abandon the Venutian viewpoint and proceed to study the theory of gravitation from the terrestrian point of view developed by Einstein.

6.3   THE CONSTRUCTION OF INVARIANTS WITH RESPECT TO INFINITESIMAL TRANSFORMATIONS

To solve the problem of constructing solutions to satisfy eq.(6.2.3) we shall convert that equation to an equivalent statement of a property of F. We first note that eq.(6.2.3) is a vector equation. If we take the dot product of the equation with an arbitrary vector $A^\lambda(x)$, and integrate over all space, we deduce an equation which looks a little different:

$$\int d\tau\left[A^\lambda(x)g_{\sigma\lambda}(x)\left(\frac{\delta F}{\delta h_{\sigma\nu}}\right)_{,\nu} + A^\lambda(x)[\sigma\nu,\lambda]\left(\frac{\delta F}{\delta h_{\sigma\nu}}\right)\right] = 0. \qquad (6.3.1)$$

If F satisfies this for arbitrary A_λ, then eq.(6.2.3) is implied. We may now integrate by parts the first term in the integrand, so as to get rid of the gradient with respect to ν. We deduce that

$$\int \underline{d\tau}\left(\frac{\delta F}{\delta h_{\sigma\nu}}\right)\left[-\left(A^\lambda(x)g_{\lambda\sigma}(x)\right)_{,\nu} + [\sigma\nu,\lambda]A^\lambda(x)\right] = 0. \qquad (6.3.2)$$

We have put a bar under $d\tau$ to remind us that we are to take the average of this integral and the corresponding integral having σ and ν interchanged: since $h_{\sigma\nu}$ is symmetric, a meaningful mathematical identity is obtained only in the case that the bracket is also symmetric in σ and ν. We can interpret this eq.(6.3.2) in another way. We note that if we make a first order change in h, say let $h_{\sigma\nu}$ change to $h_{\sigma\nu} + \xi_{\sigma\nu}$, the value of F changes as follows:

$$F[h_{\sigma\nu} + \xi_{\sigma\nu}] = F[h_{\sigma\nu}] + \xi_{\sigma\nu} \frac{\delta F}{\delta h_{\sigma\nu}} + \dots . \tag{6.3.3}$$

Therefore, our equation (6.3.2) tells us that for $\xi_{\sigma\nu}$ infinitesimal, and of the form appearing in eq.(6.3.2), F remains unchanged.

Let the field tensor $h_{\mu\nu}$ be changed by an infinitesimal transformation A^λ into the tensor $h'_{\mu\nu}$. We express $h'_{\mu\nu}$ according to the rule implied in eq.(6.3.2) as follows (we must remember to symmetrize in $\sigma\nu$ and to use the explicit expression for $[\sigma\nu, \lambda]$):

$$h'_{\sigma\nu} = h_{\sigma\nu} - \frac{1}{2} g_{\sigma\nu,\lambda} A^\lambda - g_{\lambda\sigma} A^\lambda{}_{,\nu}. \tag{6.3.4}$$

For convenience we let $-\lambda A^\nu = \zeta^\nu$ and write the equation in terms of $g_{\sigma\nu}$ instead of $h_{\sigma\nu}$ as follows:

$$g'_{\sigma\nu} = g_{\sigma\nu} + g_{\sigma\lambda}\zeta^\lambda{}_{,\nu} + g_{\nu\lambda}\zeta^\lambda{}_{,\sigma} + \zeta^\lambda g_{\sigma\nu,\lambda}. \tag{6.3.5}$$

Our problem then is this: To find a form for F, a functional of $g_{\mu\nu}$ such that under the infinitesimal transformation, eq.(6.3.5), of the $g_{\mu\nu}$ to $g'_{\mu\nu}$, the F is unchanged to first order in ζ^λ for any $\zeta^\lambda(x)$. Methods for attack for equations similar to this have been developed by mathematicians* working in differential geometry (in fact this very problem is solved in differential geometry), so we shall assume as well-educated Venutian physicists, that books giving us hints on how to proceed are available.

In fact, it might be recognized that the transformation, eq.(6.3.5), is the transformation of a tensor field $g_{\mu\nu}(x)$ under an infinitesimal transformation of coordinates $x^\lambda = x'^\lambda + \zeta^\lambda$. However, we shall continue to play our game, and try to continue to derive our results as Venutians unaware of any geometrical interpretation. Of course, we shall come back and discuss this geometrical view when we discuss Einstein's point of view.

We now proceed to find the desired invariant form for F. We shall find it useful to define a matrix which is the reciprocal of $g_{\mu\nu}$, using superscripts as indices rather than subscripts.

$$g^{\mu\nu} g_{\nu\sigma} = \delta^\mu_\sigma. \tag{6.3.6}$$

The symbol δ^μ_σ is now a true Kronecker delta, which is 1 if $\mu = \sigma$, and zero if $\mu \neq \sigma$.

---

* See for example, [Vebl 27].

The reciprocal of a matrix $A' = A + B$, if B is infinitesimal, is given by the expansion

$$\frac{1}{A'} = \frac{1}{A} - \frac{1}{A} B \frac{1}{A} + \frac{1}{A} B \frac{1}{A} B \frac{1}{A} - \dots. \tag{6.3.7}$$

Since ζ^λ is infinitesimal, we may easily construct the reciprocal of $g'_{\sigma\nu}$ according to the rule, eq.(6.3.7).

$$g'^{\alpha\beta} = g^{\alpha\beta} - \zeta^\alpha_{,\nu} g^{\nu\beta} - \zeta^\beta_{,\nu} g^{\nu\alpha} - \zeta^\lambda g^{\alpha\sigma} g^{\beta\nu} g_{\sigma\nu,\lambda} + \dots \tag{6.3.8}$$

We now examine briefly one invariant which may easily be found, in order to understand the methods, and in the next Section construct a more complicated invariant which leads us to our complete theory.

Let us consider how the determinant of a matrix changes as we change the matrix a little. We use the following expression for the determinant:

$$\text{Det } A = e^{\text{Tr} \log A}. \tag{6.3.9}$$

We shall not stop here to discuss the proof of such an equality; to make it appear reasonable, we might notice that it is trivially true in the case that the matrix is in diagonal form.

$$\text{Det } A = A_{11} A_{22} A_{33} \cdots = e^{\log A_{11} + \log A_{22} \cdots} = e^{\text{Tr} \log A} \tag{6.3.10}$$

We now apply the rule, eq.(6.3.9), to compute the determinant of $(A+B)$, where B is an infinitesimal matrix. What we need is the matrix logarithm of $A + B$; the proper expansion is

$$\text{Det}\left[A \left(1 + \frac{1}{A} B \right) \right] = \text{Det } A \cdot \text{Det} \left(1 + \frac{1}{A} B \right) = \text{Det } A \, e^{\text{Tr}(\log(1 + \frac{1}{A} B))}$$

$$= \text{Det } A \, e^{\text{Tr} \frac{1}{A} B}. \tag{6.3.11}$$

We now use this rule to compute the determinant of $g'_{\sigma\nu}$ and take the logarithm of the resulting expression.

$$\log(-\text{Det } g') = \log(-\text{Det } g) + 2\zeta^\lambda_{,\lambda} + \zeta^\lambda g_{\sigma\nu,\lambda} g^{\sigma\nu} \tag{6.3.12}$$

The product of g matrices in the last term may be related to the determinant as follows.

$$g_{\sigma\nu,\lambda} g^{\sigma\nu} = [\log(-\text{Det } g)]_{,\lambda} \tag{6.3.13}$$

What we have achieved is a new expression involving ζ^λ and its gradients together with numbers, not matrices. We set $C = \log(-\text{Det } g)$ and rewrite the resulting equation as

$$C' = C + 2\zeta^\lambda_{,\lambda} + C_{,\lambda} \zeta^\lambda. \tag{6.3.14}$$

If this expression were a pure derivative, we could integrate over all space to obtain our invariant. The form of the last two terms suggest $\exp(C/2)$ as an integrating factor. Therefore, we search for an invariant of the form $\exp(\alpha C')$, adjusting α properly. Since ζ^λ is infinitesimal, an expansion keeping the first term only gives

$$e^{\alpha C'} = e^{\alpha(C+2\zeta^\lambda,_\lambda+C,_\lambda\zeta^\lambda)} = e^{\alpha C} + e^{\alpha C}\left(\alpha 2\zeta^\lambda,_\lambda + \alpha C,_\lambda\zeta^\lambda\right). \quad (6.3.15)$$

The second term of this expression has a form which may be converted to a pure derivative; we note that

$$\left(e^{\alpha C}\zeta^\lambda\right)_{,\lambda} = e^{\alpha C}\zeta^\lambda,_\lambda + \alpha C,_\lambda\zeta^\lambda e^{\alpha C}, \quad (6.3.16)$$

which is the same as the second term of eq.(6.3.15) for $\alpha = 1/2$. When we integrate eq.(6.3.15) over all space, for $\alpha = 1/2$ the integral of the second term vanishes, and we are left with the equality

$$\int d\tau\, \exp(C'/2) = \int d\tau\, \exp(C/2). \quad (6.3.17)$$

The invariant solution, in terms of the matrix $g_{\sigma\nu}$, is therefore

$$^\circ F = \int d\tau\, \sqrt{-\mathrm{Det}\, g}. \quad (6.3.18)$$

6.4   THE LAGRANGIAN OF THE THEORY CORRECT TO ALL ORDERS

The invariant $^\circ F$ of the previous section is indeed a solution of the differential functional equation (6.2.3), but it is not the solution needed for our theory, since it involves no derivatives. In this section we shall construct the solution needed by our theory by an analogous method. The success of the trickery hinges on finding a perfect divergence which can be integrated over all space.

The starting point is again the equation (6.3.5) involving ζ_λ and first derivatives. The principle is this—we wish to find combinations of $g_{\mu\nu}$'s and derivatives which involve no ζ (or at least a perfect differential) when transformed. We have in eq.(6.3.5) first derivatives of ζ. If we form $g'_{\mu\nu,\sigma}$, second derivatives like $\zeta^\lambda,_{\sigma\nu}$ appear, etc. It looks like ever increasing complexity. But if the highest derivative is $\zeta^\lambda,_{\sigma\nu}$ and it occurs in only one term, isolated, we can eliminate it by subtracting the term with σ, ν reversed. (Actually in our case this won't do—the expression for $\zeta^\lambda,_{\sigma\nu}$ is itself automatically symmetrical, but we do the same trick at a higher derivative.) To begin then we form $g'_{\mu\nu,\sigma}$ which yields second derivatives of ζ of form $\zeta^\lambda{}_{,\sigma\nu}$, but there are two of them, $\zeta^\lambda{}_{,\nu\sigma}$ and $\zeta^\lambda{}_{,\mu\sigma}$. We try to get it to one by combining it with other derivatives like $g'_{\mu\sigma,\nu}$. It turns out that we can get rid of 2 terms, but an equal number of new terms appears, so no simplification is achieved. But when we consider the third possible ordering, $g'_{\sigma\nu,\mu}$, then we obtain by additions and subtractions a new equation in which two of the terms can be added, because they are the same. One difficulty is that as we take derivatives of products the number of terms increases so much, for example

$$g'_{\mu\nu,\sigma} = g_{\mu\nu,\sigma} + g_{\mu\lambda,\sigma}\zeta^\lambda{}_{,\nu} + g_{\nu\lambda,\sigma}\zeta^\lambda{}_{,\mu} + g_{\mu\lambda}\zeta^\lambda{}_{,\nu\sigma} + g_{\nu\lambda}\zeta^\lambda{}_{,\mu\sigma}$$
$$+ \zeta^\lambda g_{\mu\nu,\lambda\sigma} + \zeta^\lambda{}_{,\sigma}g_{\mu\nu,\lambda}, \quad (6.4.1)$$

but when all is added and subtracted and all indices flipped, and symmetrized we find

$$[\mu\nu,\sigma]' = [\mu\nu,\sigma] + [\mu\lambda,\sigma]\zeta^\lambda{}_{,\nu} + [\nu\lambda,\sigma]\zeta^\lambda{}_{,\mu}$$
$$+ [\mu\nu,\lambda]\zeta^\lambda{}_{,\sigma} + \zeta^\lambda[\mu\nu,\sigma]_{,\lambda} - g_{\sigma\lambda}\zeta^\lambda{}_{,\mu\nu}, \quad (6.4.2)$$

so only one $\zeta^\lambda{}_{,\mu\nu}$ appears. We must now get the $g_{\sigma\lambda}$ away by multiplying by the reciprocal matrix. First we introduce a new notation that simplifies the handling. We let

$$g^{\tau\sigma}[\mu\nu,\sigma] = \Gamma^\tau_{\mu\nu}. \quad (6.4.3)$$

If we multiply eq.(6.4.2) by $g^{\sigma\tau}$ in order to isolate the remaining second derivative, the equation becomes, in terms of the new symbols (called holonomic connection coefficients or Christoffel symbols)

$$\Gamma^\tau_{\mu\nu}{}' = \Gamma^\tau_{\mu\nu} + \Gamma^\tau_{\mu\lambda}\zeta^\lambda{}_{,\nu} + \Gamma^\tau_{\nu\lambda}\zeta^\lambda{}_{,\mu} - \Gamma^\lambda_{\mu\nu}\zeta^\tau{}_{,\lambda} + \Gamma^\tau_{\mu\nu,\lambda}\zeta^\lambda - \zeta^\tau{}_{,\mu\nu}. \quad (6.4.4)$$

This is automatically symmetric in $\mu\nu$. To go further we differentiate again. If we differentiate this equation with respect to a new index ρ, and subtract the corresponding equation having ρ and ν interchanged, only the following terms are not cancelled in the subtraction

$$\Gamma^\tau_{\mu\nu,\rho}{}' - \Gamma^\tau_{\mu\rho,\nu}{}' = \Gamma^\tau_{\mu\nu,\rho} + \Gamma^\tau_{\mu\lambda,\rho}\zeta^\lambda{}_{,\nu} + \Gamma^\tau_{\nu\lambda,\rho}\zeta^\lambda{}_{,\mu} + \Gamma^\tau_{\mu\nu,\rho\lambda}\zeta^\lambda$$
$$+ \Gamma^\tau_{\rho\lambda}\zeta^\lambda{}_{,\mu\nu} - \Gamma^\lambda_{\mu\nu}\zeta^\tau{}_{,\lambda\rho} \quad \text{minus } \nu, \rho \text{ reversed.} \quad (6.4.5)$$

The trick is now to get rid of two double derivatives. These come multiplied by Γ's. But in eq.(6.4.4) we have an expression which yields just $\zeta^\lambda{}_{,\mu\nu}$. We use it to supply counter terms to those in eq.(6.4.5). This can be accomplished by taking the product of two equations such as eq.(6.4.4); we see that the indices in the Γ's of one term are the same as those of the ζ in the *other* term; so that by taking a product of two (6.4.4) equations, one having the set of indices $(\tau\rho\lambda)$, the other $(\lambda\mu\nu)$, replacing (τ, μ, ν) and adding to eq.(6.4.5), the second derivatives cancel.

We introduce a new quantity $R^\tau{}_{\mu\nu\rho}$ defined as follows

$$R^\tau{}_{\mu\nu\rho} = \Gamma^\tau_{\mu\nu,\rho} + \Gamma^\tau_{\rho\lambda}\Gamma^\lambda_{\mu\nu} - \Gamma^\tau_{\mu\rho,\nu} - \Gamma^\tau_{\nu\lambda}\Gamma^\lambda_{\mu\rho}. \qquad (6.4.6)$$

Note that this tensor is explicitly antisymmetric in ρ and ν. In terms of this, the equation we finally obtain is

$$R'^\tau{}_{\mu\nu\rho} = R^\tau_{\mu\nu\rho} + \zeta^\lambda{}_{,\nu}R^\tau_{\mu\lambda\rho} + \zeta^\lambda{}_{,\rho}R^\tau_{\mu\nu\lambda} + \zeta^\lambda{}_{,\mu}R^\tau_{\lambda\nu\rho} + \zeta^\tau{}_{,\lambda}R^\lambda_{\mu\rho\nu} + \zeta^\lambda R^\tau_{\mu\nu\rho,\lambda}.$$
$$(6.4.7)$$

What must now be done is to treat this equation as we treated eq.(6.3.5), which is of the same form, except that it is the tensor $R^\tau{}_{\mu\nu\rho}$ which is involved rather than $g_{\sigma\nu}$. A procedure entirely analogous to the one we used before leads to the answer for the invariant quantity F:

$$F = -\frac{1}{2\lambda^2} \int d\tau \, g^{\mu\nu} \, R^\tau{}_{\mu\nu\tau} \sqrt{-\mathrm{Det}\, g_{\alpha\beta}}. \qquad (6.4.8)$$

This is the quantity we were after, part of the action of a theory correct to all orders.

6.5 THE EINSTEIN EQUATION FOR THE STRESS-ENERGY TENSOR

The functional F which we have just deduced results in a Venutian theory of gravitation which is identical to that developed by Einstein. If we make an expansion of the functional F when the gravitational fields are weak, we obtain as the leading terms the F^2 and F^3 functionals of our earlier theory. We may say therefore that our Venutian viewpoint has succeeded in its aim to construct a self-consistent theory of gravitation by means of successive logical steps guessed at by analogy but without apparently demanding a superhumanly keen intuition.

Einstein himself, of course, arrived at the same Lagrangian but without the help of a developed field theory, and I must admit that I have no idea of how he ever guessed at the final result. We have had troubles enough in arriving at the theory—but I feel as though he had done it while swimming underwater, blindfolded, and with his hands tied behind his back! Nevertheless, now that we have arrived at an equivalent theory, we shall abandon the Venutian viewpoint and discuss the terrestrian point of view due to Einstein.

We shall use the following standard notation for three tensors deduced from our $R^\tau{}_{\mu\nu\rho}$ by multiplying by $g_{\alpha\beta}$ and contracting:

$$g_{\tau\sigma} R^\tau{}_{\mu\nu\rho} = R_{\sigma\mu\nu\rho}, \qquad \text{Antisymmetric in } (\sigma\mu) \text{ and } (\nu\rho)$$
$$R^\tau{}_{\mu\nu\tau} = R_{\mu\nu}, \qquad\qquad\qquad\qquad\qquad\qquad (6.5.1)$$
$$g^{\mu\nu} R_{\mu\nu} = R.$$

The quantity $R_{\sigma\mu\nu\rho}$ is a tensor (the Riemann tensor). It is antisymmetric for an interchange of ν and ρ, also antisymmetric for interchange of $\sigma\mu$, and symmetric if the pair $\sigma\mu$ is interchanged with the pair $\nu\rho$. $R_{\mu\nu}$ (the Ricci tensor) is symmetric.

The variation of the functional F, eq.(6.4.8), with respect to $g_{\mu\nu}$ yields

$$2\frac{\delta F}{\delta g_{\mu\nu}} = -\frac{1}{\lambda^2} \frac{\delta(\sqrt{-g}R)}{\delta g_{\mu\nu}} = \frac{1}{\lambda^2} \sqrt{-g} \left(R^{\mu\nu} - \frac{1}{2} g^{\mu\nu} R \right), \qquad (6.5.2)$$

where

$$g \equiv \text{Det } g_{\mu\nu}.$$

The last quantity in eq.(6.5.2) is the stress-energy tensor of our theory (see eq.(6.2.2)), and it satisfies the following equation

$$g_{\sigma\lambda} T^{\lambda\nu}{}_{,\nu} = -[\mu\nu, \sigma] T^{\mu\nu} \qquad \left(\text{or} \quad T^{\lambda\nu}{}_{,\nu} = -\Gamma^\lambda_{\mu\nu} T^{\mu\nu} \right), \qquad (6.5.3)$$

if substituted for $T^{\mu\nu}$ as we required it to do. That is, the full equations of the gravitational field to all orders are

$$\sqrt{-g} \left(R^{\mu\nu} - \frac{1}{2} g^{\mu\nu} R \right) = \lambda^2 T^{\mu\nu}, \qquad (6.5.4)$$

where $T^{\mu\nu}$ is our matter energy tensor. This is the equation Einstein obtained.

7.1 THE PRINCIPLE OF EQUIVALENCE

As our next project, we shall describe relativity and gravitation from a point of view which is more nearly in accord with the approach of Einstein. We hope that by viewing the theory from different vantage points, we increase our understanding of the theory as a whole. The theory of gravitation as viewed within the framework of Einstein's ideas is something so beautifully exciting that we shall be sorely tempted to try to make all other fields look like gravity, rather than continue with the Venutian trend of making gravity look like other fields that are familiar to us. We shall resist the temptation.

The seeds for Einstein's approach are to be found in the physics known in his day, electrodynamics and Newtonian mechanics. One feels that the idea uppermost in Einstein's mind as he developed his theories was that all branches of physics should be consistent; he had found a way to reconcile the Lorentz invariance of classical electrodynamics with the apparent Galilean invariance of Newtonian mechanics, and many new physical results followed. Similarly, it was a puzzling fact about gravity that led to his theory of gravitation when he converted this fact into a physical principle.

The central idea of gravity, the most cogent fact about how it acts, is that weight and mass are exactly proportional, so that all objects accelerate under gravity at exactly the same rate, no matter what their constitution may be. The experiment of Eötvös showed how a centrifugal force added to a gravity force in such a way that the resultant was indistinguishable from a purely gravitational effect. These facts suggested to Einstein that perhaps there might be a physical principle that made accelerations imitate gravity in all respects. It is quite obvious that mechanical experiments performed inside of an accelerating box yield results which are indistinguishable from those which would be obtained if the box were stationary, but in the presence of a gravitational field. There were no direct verifications of this in Einstein's day, but today we are familiar with the weightlessness in satellites, which is a cancellation of gravitational forces by an acceleration. It is this possibility of cancellation which is the core of the principle of equivalence.

Before we can get useful physics out of the idea, we must have a statement which is more precise, and which involves definite measurable quantities. A more precise statement which makes good sense in terms of Newtonian mechanics might involve forces on stationary objects. If we accelerate a box with a uniform acceleration g, on all objects which are at rest inside the box, there must be a force which is exactly proportional to the weight; for example, the stresses on the tables or springs supporting the weights inside the box are the same as those which would be produced by a uniform gravity field of strength g. Now, since the stresses on tables are not directly measurable quantities, it is perhaps better to think of weights hanging by springs. The displacements from the unstressed lengths for given masses hanging from given springs should be exactly equal when (1) the box is stationary in a gravitational field of strength g, or (2) the box is being accelerated with a uniform rate g in a region of zero field, as illustrated in Figure 7.1

This statement of the principle of equivalence is more physical, but we are still talking in an elementary level without defining the nature of forces very precisely. Is it possible to make meaningful physical statements without defining the nature of forces? We may recall the situation in Newtonian mechanics. It is often said that Newton's law,

$$F_x = m\ddot{x}, \tag{7.1.1}$$

is simply a definition of forces, so that it has no real physics—that it involves circular reasoning. But evidently the whole theory of Newton is not circular, since it predicts correctly the orbits of the moon and of the planets. What Newton meant by telling us that we should calculate forces according to the rule eq.(7.1.1) was that, if there exists an acceleration, we are to look about for something physical which produces such a force. The future of physics lies in finding out how the environment of an object

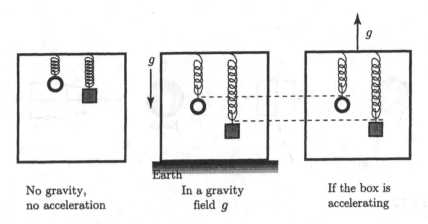

| No gravity, | In a gravity | If the box is |
| no acceleration | field g | accelerating |

Figure 7.1

is connected to the forces that we put in the left-hand side of the equation
(7.1.1) to match the observed accelerations.

When Newton goes on to his third law,

$$F(\text{action}) = -F(\text{reaction}), \qquad (7.1.2)$$

he is making a physical statement, since he is making a specification on
the connection between forces and physical objects. His law of gravity
is another specification of how the environment of an object is related
to its accelerations. The second law is given in the spirit of *"cherchez la
femme"*: If we see a force, we are to search for the guilty object which is
producing it.

In a similar way, our simple formulation of the principle of equivalence
makes a physical statement about how environments affect things; it does
not depend on the correctness of Newton's Second Law, eq.(7.1.1). The
environment in this case consists either in the masses which produce
gravitational fields, or in the accelerations.

It is not possible to cancel out gravity effects entirely by uniform
accelerations. We imagine a box in orbit about the earth, a satellite. Since
the earth's field is not uniform, it is only at one point, near the center
of mass of the satellite, that gravitational effects are exactly balanced by
the acceleration. As we go far away from the center of mass, the earth's
field changes either strength or direction, so that there will be small
uncancelled components of the gravitational forces. If the box is not very
large, these small additional forces are very nearly proportional to the
distance from the center of the box, and they have a quadrupole character
as illustrated in Figure 7.2(a). Forces such as these cause tides on the
earth, so we may call them the tidal forces. We may also consider the box

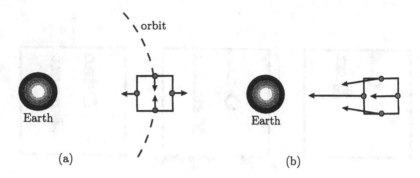

Figure 7.2

to be in a similar nonuniform field, but unaccelerated as illustrated in Figure 7.2(b). The principle of equivalence now tells us that we can create a situation physically indistinguishable from that inside the satellite, if we put large masses far away so that we superpose a uniform field which exactly cancels out gravity at the center of the box.

We begin to see how we might make yet a better statement of equivalence: one gravitational field within an accelerated box is equivalent to a different gravitational field and a different acceleration of the box. We can get rid of gravity at any one point and at any one time; over a small region about such a point, the residual differences should be proportional to the distance from the point of cancellation. It becomes obvious that in constructing our theory we will be considering transformations that we may symbolize as

$$(\text{gravity})' = (\text{gravity}) + (\text{acceleration}). \qquad (7.1.3)$$

Because of this possibility, we shall not in any absolute way be able to say that one effect is gravitational and one is inertial; it will not be possible to define a "true" gravity, since we cannot ever define precisely how much of an observed force is given by gravity and how much is due to an acceleration. It is true that we cannot imitate gravity with accelerations everywhere, that is, if we consider boxes of large dimensions. However, by considering the transformations, eq.(7.1.3), over infinitesimal regions, we expect to learn how to describe the situation in differential form; only then shall we worry about boundary conditions, or the description of gravity over large regions of space.

In special relativity, extensive use is made of reference frames which are moving with a uniform velocity in a straight line. But, as soon as we allow the presence of gravitating masses anywhere in the universe, the concept of such truly unaccelerated motion becomes impossible, because there will be gravitational fields everywhere.

If we are performing experiments inside of a box which is not in free fall, it will be possible to detect the presence of gravitation-like forces, by the experiments with springs, for example. However, we cannot tell from inside the box whether we are accelerating relative to the nebulae, or whether the forces are due to masses in the neighborhood. It is this salient fact about gravitating forces which gives the hint for a postulate which eventually leads us to the complete theory.

We postulate: It shall be impossible, by any experiment whatsoever performed inside such a box, to detect a difference between an acceleration relative to the nebulae and gravity. That is, an accelerating box in some gravitational field is indistinguishable from a stationary box in some different gravitational field.

How much like Einstein this sounds, how reminiscent of his postulate of special relativity! We know the principle of equivalence works for springs, (as we knew special relativity worked for electrodynamics), and we extend it by *fiat* to all experiments whatsoever. We are used to such procedures by now, but how originally brilliant it was in 1911—what a brilliant, marvelous man Einstein was!

7.2 SOME CONSEQUENCES OF THE PRINCIPLE OF EQUIVALENCE

The principle of equivalence tells us that light falls in a gravitational field. The amount by which light should fall as it travels a given distance in a region of uniform gravitational field may be very simply calculated by thinking of the accelerating box; if the box is accelerating, the light travels in a straight line in an unaccelerated system, hence it is simple kinematics to calculate the path of the light inside the box. The experiment needs only a light source and a detector, and a series of slits to define the path of the light as illustrated in Figure 7.3.

We cannot use such simple means to compute the deflection of light by a star, because the field of a star is not uniform; a similar simple-minded calculation would be wrong by a factor of two if we simply use Newtonian potentials; the correct calculation needs to use appropriate relativistic fields.

The principle of equivalence also tells us that clock rates are affected by gravity. Light which is emitted from the top of the accelerating box will look violet-shifted as we look at it from the bottom. Let us do some calculations, appropriate for small velocities. The time that light takes to travel down is to a first approximation c/h, where h is the height of the box in Figure 7.3. In this time, the bottom of the box has acquired a

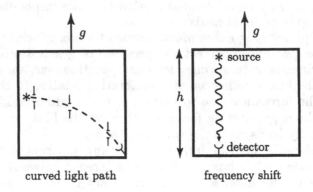

curved light path   frequency shift

Figure 7.3

small additional velocity, $v = gh/c$. The net effect is that the receiver is moving relative to the emitter, so that the frequency is shifted

$$\begin{aligned} f_{\text{received}} &= f_{\text{emitted}}(1 + v/c) \\ &= f_{\text{emitted}}(1 + gh/c^2). \end{aligned} \qquad (7.2.1)$$

Thus, the clock at the bottom will receive a frequency different from the emitted frequency. Note that this conclusion does not depend on $E = mc^2$ and on the existence of energy levels, which we had to postulate in the argument we have previously given. The conclusion is based on the expected behavior of classical objects; the calculation results from geometry and kinematics, and makes a direct physical prediction from the postulate of equivalence. As before, there is no paradox; the clock at the top looks bluer, a man living at the top of the box looks bluer than the man living at the bottom. Similarly, we may compute the frequency shift for light emitted by the man living downstairs. Since the receiver is receding from the source in this case, the man downstairs looks redder when viewed from upstairs.

One of the ways of describing this situation is to say that the time scale is faster at the top; time flows are different in different gravitational potentials, so that time flows are unequal in various parts of the world. How much is this time difference at various points in space? To calculate it, we compare the time rates with an absolute time separation, defined in terms of the proper times ds. Let us suppose that there are two events occurring at the top, which are reported to be a time dt apart; then

$$\phi = gh; \qquad ds = dt\left(1 + \phi/c^2\right), \qquad (7.2.2)$$

in the limit of small velocities. The quantity ϕ is simply the potential difference between the location of the events, and the point of reference. A more careful computation gives us an expression good for all velocities:

$$ds = dt \sqrt{1 + 2\phi/c^2}. \tag{7.2.3}$$

Again, we should recall that we cannot simply use the Newtonian potentials in this expression; our definition of ϕ must be relativistically accurate.

7.3   MAXIMUM CLOCK RATES IN GRAVITY FIELDS

Now that we have concluded that gravitational effects make clocks run faster in regions of higher potential, we may pose an amusing puzzle. We know that a clock should run faster as we move it up, away from the surface of the earth. On the other hand, as we move it, it should lose time because of the time dilation of special relativity. The question is, how should we move it up or down near the surface of the earth so as to make it gain as much time as possible? For simplicity, we consider the earth to have a uniform field, and consider motion in only one dimension. The problem clearly has a solution. If we move very fast, at the speed of light, the clock does not advance at all, and we get behind earth time. If we move the clock up a very small distance, and hold it there, it will gain over earth time. It is clear that there is some optimum way of moving the clock, so that it will gain the most in a given interval of earth time. The rules are that we are to bring the clock back to compare it with the stationary clock.

We shall give the answer immediately, although it would be a good exercise to work it out in detail. To make the moving clock advance the most, in a given interval of ground time, say 1 hour, we must shoot it up at such a speed that it is freely falling all the time, and arrives back just in time, one hour later. See Figure 7.4. The problem is more difficult if we try to do it in more dimensions, but the same answer is obtained; if we want to make the clock return one hour later, but to a different spot on earth, we must shoot the clock up into the ballistic trajectory. The same answer is obtained in a nonuniform gravitational field. If we are to shoot a clock from one earth satellite to another, the true orbit is precisely that which gives the maximum proper time.

In working out these problems, some troubles arise because we have not made accurate definitions. For example, the free-fall solutions are not necessarily unique; the ballistic "cannonball" problem in general has two solutions, that is, two angles and initial velocities will give maxima (the satellite orbit may go the long way about the earth). Nevertheless, any of these solutions correspond to maxima of the elapsed time for the moving

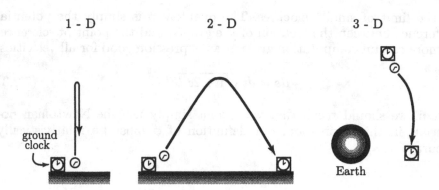

Figure 7.4

clock. Whether they are relative or absolute maxima is not so important for our purposes—what is important is that these solutions suggest how we may obtain mechanics from a variational principle.

To understand the significance of the maximal property of elapsed times, we may consider what happens in the limit of small velocities. The elapsed time is an integral of ds, which represents the ticking rate of the clock. In the nonrelativistic limit, the integral to be maximized is

$$\int ds = \int_{t_0}^{t_1} dt \left(1 + \frac{\phi}{c^2} - \frac{v^2}{2c^2} + \dots \right), \qquad (7.3.1)$$

as

$$\sqrt{1 - v^2/c^2} \approx 1 - \frac{v^2}{2c^2}.$$

The first term integrates to the time difference in the reference system, $(t_1 - t_0)$. The other two terms may be rearranged to look like something which should be very familiar, by multiplying by the mass of the particle and changing sign,

$$\int ds = (t_1 - t_0) - \frac{1}{mc^2} \int_{t_0}^{t_1} dt \left(\frac{1}{2} mv^2 - m\phi \right). \qquad (7.3.2)$$

To maximize this expression, for a fixed time interval $(t_1 - t_0)$, we take the minimum of the integral on the right. But this integral is nothing but the classical action for a particle of mass m in a gravitational potential ϕ. We see that the requirement that the proper time should be a maximum is equivalent to the principle of least action in the classical limit.

These results suggest how we might obtain a mechanical principle (equivalent roughly to Newton's Second Law) which will be relativistic. It is that the variation of $\int ds$ should be zero.

$$\delta \int_{1}^{2} ds = 0 \qquad (7.3.3)$$

It was Einstein's guess that this principle will give the motion in the presence of the gravitational fields. This solved the problem of finding the equations of motion given the field. The remaining problem is now to connect the potential ϕ which appears in this expression with the environment. This was the great problem before Einstein. How do we get the correct ϕ? What happens if we use the wrong gravity, as if we worked in a system in which there were centrifugal forces, but did not know it? We have seen that the gravitational forces are inextricably mixed with inertial forces, so that we cannot make a universally valid separation of the two.

It was Einstein's guess that in situations such as these, it should not matter whether we use the universally correct ϕ or not; if this ϕ is correctly defined, the description of the physics should be independent of the particular way in which we have separated inertial and gravitational effects. Thus, in order to construct a formula for ϕ which will have this property, what we must do is to study very carefully the way in which the proper time interval ds is expressed in different coordinate systems, as we apply transformations such as those we have symbolized by gr' = gr + accel. Such a study may allow us to construct an expression for ds which will be invariant under all possible transformations.

7.4   THE PROPER TIME IN GENERAL COORDINATES

To get Einstein's formula for $(ds)^2$ we must consider reference systems which are not only accelerating but also are in the process of being distorted in an arbitrary fashion. We want a general formula for the coordinates which is analogous to the specification of rotating coordinates.

$$x' = x \cos \omega t + y \sin \omega t \qquad z' = z$$
$$y' = y \cos \omega t - x \sin \omega t \qquad t' = t \qquad (7.4.1)$$

We describe a general acceleration and stretching by specifying how each of the four coordinates in one system depends on all the coordinates of another system.

$$x = x(x', y', z', t') \qquad z = z(x', y', z', t')$$
$$y = y(x', y', z', t') \qquad t = t(x', y', z', t') \qquad (7.4.2)$$

Let us first consider what the situation is when $\phi = 0$. In this case, we know that the proper time in the untwisted system is simply (here we set $c = 1$)

$$(ds)^2 = (dt)^2 - (dx)^2 - (dy)^2 - (dz)^2. \qquad (7.4.3)$$

To describe the proper time in the primed coordinates, we simply rewrite the differentials as follows:

$$dx^\mu = \frac{\partial x^\mu}{\partial x'^\alpha} dx'^\alpha, \qquad (ds)^2 = \eta_{\mu\nu} dx^\mu dx^\nu. \qquad (7.4.4)$$

This defines a metric tensor $g'_{\alpha\beta}$ which contains the description of the arc length ds in the arbitrarily twisting and accelerating system.

$$(ds)^2 = \eta_{\mu\nu} \frac{\partial x^\mu}{\partial x'^\alpha} \frac{\partial x^\nu}{\partial x'^\beta} dx'^\alpha dx'^\beta = g'_{\alpha\beta} dx'^\alpha dx'^\beta \qquad (7.4.5)$$

We note that $g'_{\alpha\beta}$ represents ten functions of the coordinates (x', y', z', t'), since there are ten bilinear products $dx^\alpha dx^\beta$. The metric tensor is symmetric. Once we have these ten functions specified, it should be a purely mathematical exercise to find the paths which will make the proper time a maximum.

What will happen when the gravity is not zero? In the simple cases that we considered in the last section, we found that the proper time was given by something like

$$(ds)^2 = (1 + 2\phi/c^2)(dt)^2 - (dx)^2 - (dy)^2 - (dz)^2. \qquad (7.4.6)$$

This is only slightly different from the case of zero field. It was Einstein's idea that the complete description of gravity could always be specified by a metric tensor $g_{\alpha\beta}$ such that

$$(ds)^2 = g_{\alpha\beta} dx^\alpha dx^\beta. \qquad (7.4.7)$$

The case of zero field corresponds to a particularly simple form for the metric tensor, $g_{\alpha\beta} = \eta_{\alpha\beta}$. As we change the coordinate system, the new metric tensor is given by

$$g'_{\alpha\beta} = \frac{\partial x^\mu}{\partial x'^\alpha} \frac{\partial x^\nu}{\partial x'^\beta} g_{\mu\nu}. \qquad (7.4.8)$$

As before, the motion of particles is given by the requirement that the proper time is to be a maximum. If it is possible, by some kind of judicious choice of transformation, to reduce the tensor so that $g'_{\alpha\beta} = \eta_{\alpha\beta}$, then we may conclude that there is no gravitational field, and that there is no acceleration. But this cannot be done in general, since the general tensor $g_{\alpha\beta}$ represents ten presumably independent functions, and only four functions can be specified in the transformation of coordinates, eq.(7.4.2). Only under very special circumstances can accelerations transform away all the $g_{\mu\nu}$ everywhere. If there really is some matter in the environment, then the reduction to $\eta_{\alpha\beta}$ will be impossible. In that case all possible tensors $g_{\alpha\beta}$ related by eq.(7.4.8) will be equivalent, since none of them will lead to particularly simple expressions for $(ds)^2$.

What we have achieved is to learn something of the character of the description of gravitational forces. In Newtonian theory, the corresponding thing is the statement that the force is given by the gradient of a scalar function.

$$\text{Newtonian Gravity}: \quad m\ddot{x} = F_x, \qquad F_x = -\nabla\phi$$
$$\text{Einstein Theory}: \quad \delta \int ds = 0, \qquad (ds)^2 = g_{\mu\nu}dx^\mu dx^\nu \tag{7.4.9}$$

The second part of the theory corresponds to the specification of how the potentials (ϕ or $g_{\mu\nu}$) are related to matter. In the Newtonian theory we have

$$\Delta\phi = 4\pi G\rho. \tag{7.4.10}$$

We shall eventually arrive at a specification of the tensor $g_{\mu\nu}$ in terms of the matter. The central idea is that since matter is physical, whereas coordinate systems are not, matter must be described in such a way that the results of solving the equation of motion are independent of the particular choice of coordinates—the meaningful properties of the tensor $g_{\mu\nu}$ are expected to be invariant quantities under arbitrary transformations.

7.5   THE GEOMETRICAL INTERPRETATION OF THE METRIC TENSOR

The tensor $g_{\mu\nu}$ can be given a geometric interpretation. We shall study briefly its significance in the case of two dimensions, in order to help us gain insight as to what invariants are imbedded in it. In the case of the uniform gravitational fields, we have seen that the tensor $g_{\mu\nu}$ describes how the scale of time is different at different locations in space. More generally, the tensor represents how the scales vary from point to point, not only for the time, but for the space coordinates as well. In orthogonal Cartesian coordinates, the two-dimensional arc length ds is given by

$$(ds)^2 = (dx)^2 + (dy)^2. \tag{7.5.1}$$

In plane polar coordinates, the arc length is given by

$$(ds)^2 = (dr)^2 + r^2(d\theta)^2. \tag{7.5.2}$$

Evidently, there is no significance to the symbols that we use for the coordinates, and physics in a plane should be identical whether we use Cartesian axes or plane polar axes to describe it. This means that if we find that the description of the arc length in some system that we have chosen is correctly given by

$$(ds)^2 = y^2(dx)^2 + (dy)^2, \tag{7.5.3}$$

there is no deep significance to the fact that the x length seems to be varying with the coordinate y; a simple coordinate transformation restores the Cartesian form for the arc length, eq.(7.5.1).

Let us now examine a more interesting case. We imagine that we are bugs crawling along a floor of what we have always assumed were square tiles, and all our lives we have been thinking that the geometry of the floor was correctly given by counting tiles and using the Euclidean rule, eq.(7.5.1), an interval dx or dy corresponding to a number of tile lengths. But then some of the smarter bugs begin to check up on this by using rulers, and after a series of measurements come up with the result that the measured arc lengths correspond to the number of tiles as follows:

$$(ds)^2 = \frac{(dx)^2 + (dy)^2}{1 + ar^2}, \tag{7.5.4}$$

where

$$r^2 = x^2 + y^2.$$

Let us suppose that the smart bugs measured very carefully the ratio of the circumference to the radius of a circle, by laying down their rulers along curves of constant r and from the center out along one of the axes. Their results would have been

$$\text{Circumference} = \int ds = \int_0^{2\pi} \frac{r\,d\theta}{(1 + ar^2)} = \frac{2\pi r}{(1 + ar^2)}$$

$$\text{Radius} = \int_0^r ds\Big|_{y=0} = \int_0^r \frac{dx}{(1 + ax^2)} = b\,\text{Arctan}\left(\frac{r}{b}\right) = R, \tag{7.5.5}$$

where $b^2 = 1/a$.

The experimental result for the ratio of circumference to radius would have been

$$\frac{1}{R}\frac{2\pi b\tan(R/b)}{1 + \tan^2(R/b)} = 2\pi\frac{\sin(2R/b)}{(2R/b)}. \tag{7.5.6}$$

This result becomes 2π only in the limit that the radius of the circle goes to zero. It is this measurable ratio which is a significant physical result.

The particular model that we have discussed has a simple geometrical interpretation, but again we emphasize that it is the experimental results

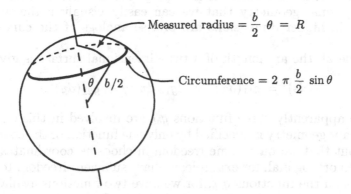

Figure 7.5

which are important, and these depend solely on the correct formula for the arc length; it would not matter at all if we could not give to the formula a geometrical sense that we can easily visualize.

We may say that all along, we bugs have been living on the surface of a sphere, without knowing it. Now that we have guessed it, we easily understand why our measurements of circumferences gave the particular result, eq.(7.5.6). If a sphere has a radius $(b/2)$, the result, eq.(7.5.6), represents the ratio of the circumference of a circle to the distance along the surface along a meridian as illustrated in Figure 7.5.

Our previous point of view about gravity may be compared to that which might be held by the more conservative of the bugs: The tiles are "really" square, but the rulers are affected as we move them from place to place, because of a certain field which has this effect. Our newer, geometrical point of view, will be that we cannot truly define the tiles to be "really" square; we live in a world which is in general not Euclidean, which has a curvature which is measurable by doing suitable experiments. There is no need to think of processes as occurring in a space which is truly Euclidean, since there is nothing physical which can ever be measured in this fictional space. The tiles represent simply a labelling of coordinates, and any other labelling would have done just as well.

7.6   CURVATURES IN TWO AND FOUR DIMENSIONS

The invariant quantities which characterize the geometry in a way which is independent of particular choices of coordinates are curvatures. It is easy to visualize the notion of curvature when we are considering a two-dimensional space: a flat, uncurved space is a plane, and a curved space is a curved surface. Although in our later work we shall need to work analytically with curvatures, it is appropriate to work a little with the two-dimensional geometry that we can easily visualize; the notions of curvature in higher dimensions are precise analogs of the curvatures of surfaces.

In general, the arc length of a two-dimensional surface is given by

$$(ds)^2 = g_{11}(dx)^2 + 2g_{12}\,dx\,dy + g_{22}(dy)^2. \qquad (7.6.1)$$

Although apparently three functions g_{ab} are involved in this expression, the invariant geometry is specified by only one function of the coordinates; it turns out that we have some freedom in choosing coordinates, we can make them orthogonal, for example; we have sufficient freedom to put two conditions on the functions g_{ab}, for we have two functions available for a coordinate transformation. In particular, it is always possible to choose

1. $g_{12} = 0$
2. $g_{22} = g_{11}.$

This means that for the purposes of studying the geometrical measurements on a two-dimensional surface, the most general form of the arc length is

$$(ds)^2 = f(x,y)\Big((dx)^2 + (dy)^2\Big). \qquad (7.6.2)$$

The function $f(x,y)$, in one viewpoint, represents the factor by which rulers are changed as we move about the surface. In the other viewpoint, it evidently determines the curvature of the surface.

An amusing example of a physical situation which corresponds exactly to these geometries was invented by one of Robertson's students. We imagine that a man makes measurements with a ruler on a hot plate, which is hotter at some places than in others. The ruler expands or contracts as it measures in hotter or cooler regions of the plate; evidently, the appropriate function $f(x,y)$ is determined by the local temperature and the coefficient of thermal expansion of the ruler.

The local curvature of a surface at a point must be defined by some mathematical criterion, involving the limiting case of measurements made on smaller and smaller objects. We might, for example, choose to compare the ratios of circumferences to radii, or the ratios of areas to the square of the radii; for the case of spherical surfaces, these ratios are different from those obtained on a plane surface by factors like $(\sin\theta)/\theta$, where θ is the ratio of the measured radius to the radius of the sphere. In the limit of smaller and smaller circles, this quantity differs from unity by an amount proportional to the area of the circle. The coefficient of proportionality (times 3) is $1/R^2$ for a sphere. This number (coefficient of area in deviation of circumference from 2π) is appropriate to describe the local curvature; it is known as the Intrinsic Curvature, or also as the

Gaussian Mean Square Curvature of the spherical surface, because the mathematics of all this is due to Gauss.

We may easily consider other curved surfaces. For example, it is easy to see that a cylindrical surface has zero curvature; since a cylindrical surface can be rolled out into a plane without stretching, it is obvious that the ratio of circumferences to radii must be exactly 2π. For more complicated cases, if the surfaces are smooth, they must look like either paraboloids or hyperbolic paraboloids over the infinitesimal regions in which we define the intrinsic curvature. These surfaces are described by two length parameters, the radii of curvature in two perpendicular planes. In this case the intrinsic curvature is given by $1/(R_1 R_2)$. It is positive if the surface is paraboloidal, or negative if the surface is hyperbolic paraboloidal. We see that this gives the correct curvatures for the special cases of spherical surfaces and cylindrical surfaces; for a sphere, both radii are equal; for a cylinder, one radius is infinite.

The curvature of a four dimensional space will be defined by analogous mathematical criteria. However, we can hardly expect to be able to construct such simple pictures in our minds, and we shall have to rely heavily on analytic methods, since our intuition will probably fail us. It is hard enough to think of the four dimensional space of Special Relativity with good intuition—I find it very difficult to visualize what is close to what, because of the minus signs. And to visualize this thing with a curvature will be harder yet. A curved two-dimensional space is conveniently thought of as a curved surface embedded in a three-dimensional space. But an analogous description for curvatures of a three-dimensional space requires conceptual embedding into six dimensions, and to do it for four dimensions, we have to think of the four dimensional space as being embedded in a ten-dimensional world. Thus, the curvatures of space-time are considerably more complicated than surface curvatures.

7.7   THE NUMBER OF QUANTITIES INVARIANT UNDER GENERAL TRANSFORMATIONS

In a four-dimensional geometry, there are some twenty coefficients which describe the curvature in a manner analogous to that in which the one quantity $1/(R_1 R_2)$ describes the intrinsic curvature of a two-dimensional surface. These twenty quantities determine the physically significant properties of the tensor $g_{\mu\nu}$; what we must do is simplify the tensor $g'_{\mu\nu}$ by a judicious choice of coordinates, in the same way that it is possible to define the geometry of two dimensions by a single function $f(x,y)$, eq.(7.6.2).

We have seen that in general we cannot get rid of gravitational fields by superposing an acceleration, except at a single point. Since curvatures must be defined by specifying what happens in an infinitesimal region about a given point, it is appropriate to study to what extent the metric tensor $g'_{\mu\nu}$ can be simplified. By analogy with the two-dimensional case, we may suspect that it is possible to choose the coordinates (called Riemann normal coordinates) in such a way that the space about the point is flat except for terms of second order in the distance about the point. In other words, a curved surface pulls away from a plane which is tangent to it by an amount which is quadratic in the coordinates measured from the point of tangency; we expect an analogous thing to occur in four dimensions.

Let us for a moment count how many quantities we can specify in the transformations, and how much we can simplify $g'_{\mu\nu}$, if we do an expansion of $g'_{\mu\nu}$ about some point x_0. Let any point in space be x, then a Taylor expansion of $g'_{\mu\nu}$ about x_0 is as follows:

$$g'_{\mu\nu}(x) = g'_{\mu\nu}(x_0) + g'_{\mu\nu,\tau}(x_0)(x^\tau - x_0^\tau) + \frac{1}{2}\, g'_{\mu\nu,\tau\sigma}(x_0)(x^\tau - x_0^\tau)(x^\sigma - x_0^\sigma) + \dots .$$
(7.7.1)

We must calculate the metric tensor $g'_{\mu\nu}(x_0)$ and its derivatives according to the rule, eq.(7.4.8); this yields:

$$g'_{\alpha\beta}(x_0) = \left[\frac{\partial x^\mu}{\partial x'^\alpha} \cdot \frac{\partial x^\nu}{\partial x'^\beta} \cdot g_{\mu\nu}\right]_{x_0},$$

$$g'_{\alpha\beta,\tau} = \left[\frac{\partial x^\mu}{\partial x'^\alpha} \cdot \frac{\partial x^\nu}{\partial x'^\beta} \cdot g_{\mu\nu,\tau}\right]_{x_0} + 2\left[\frac{\partial^2 x^\mu}{\partial x'^\alpha \partial x'^\tau} \cdot \frac{\partial x^\nu}{\partial x'^\beta} \cdot g_{\mu\nu}\right]_{x_0},$$

$$g'_{\alpha\beta,\tau\sigma} = \left[\frac{\partial x^\mu}{\partial x'^\alpha} \cdot \frac{\partial x^\nu}{\partial x'^\beta} \cdot g_{\mu\nu,\tau\sigma}\right]_{x_0} + 2\left[\frac{\partial^3 x^\mu}{\partial x'^\alpha \partial x'^\tau \partial x'^\sigma} \cdot \frac{\partial x^\nu}{\partial x'^\beta} \cdot g_{\mu\nu}\right]_{x_0}$$
$$+ \text{ other terms.}$$
(7.7.2)

We see that for the simplification of $g'_{\mu\nu}$, if we consider only an expansion to second order, we must choose our transformation so that the partial derivatives appearing in eq.(7.7.2) have certain values. We may specify the following quantities in our transformation.

1. The sixteen quantities $\left[\dfrac{\partial x^\mu}{\partial x'^\alpha}\right]_{x_0}$

2. The forty quantities $\left[\dfrac{\partial^2 x^\mu}{\partial x'^\alpha \partial x'^\beta}\right]_{x_0}$ (7.7.3)

3. The eighty quantities $\left[\dfrac{\partial^3 x^\mu}{\partial x'^\alpha \partial x'^\beta \partial x'^\gamma}\right]_{x_0}.$

(Note that the order of the derivatives makes no difference.) On the other side of the ledger, the number of values and of derivatives of the metric tensor is as follows:

1. There are 10 components $g'_{\mu\nu}(x_0)$.

2. There are forty first derivatives $g'_{\mu\nu,\tau}(x_0)$.   (7.7.4)

3. There are one hundred second derivatives $g'_{\mu\nu,\tau\sigma}(x_0)$.

We first can try to make $g'_{\mu\nu}(x_0) = \eta_{\mu\nu}$. This only involves the first derivatives $[\partial x^\mu/\partial x'^\alpha]_{x_0}$. We have 10 conditions to satisfy with 16 free parameters. We can surely do it, and have 6 degrees of freedom left over. These six are the six parameters of special relativity, Lorentz transformations and rotations (velocity vector, and rotation axis and angle) which may still be performed leaving $\eta_{\mu\nu}$ unchanged. Next we can make all 40 $g'_{\mu\nu,\tau}(x_0)$ vanish exactly by our forty

$$\left[\frac{\partial^2 x^\mu}{\partial x'^\alpha \partial x'^\beta}\right]_{x_0}.$$

The derivative of $g_{\mu\nu}$ appears in the equation of motion of minimum $\int ds$. That they can be made to vanish at a point means that all gravity forces can be removed at any one point and time by suitable accelerations, as we said.

The net result is that there remain twenty linear combinations of second derivatives of the type $g'_{\mu\nu,\tau\sigma}$ which cannot be removed by the transformation. It is these quantities which are to describe the details of the tidal forces. In the next lecture, we shall proceed to construct these twenty quantities in terms of the components of the tensor $g_{\mu\nu}$ as given in whatever coordinate system we wish to start from.

8.1 TRANSFORMATIONS OF TENSOR COMPONENTS IN NONORTHOGONAL COORDINATES

In much of the previous work, it has been possible to use a simplified notation for summations of tensor components, because we were always dealing with coordinate systems which were orthogonal. In particular, we have often used a summation rule for repeated indices.

$$A_\mu B^\mu = A_4 B_4 - A_3 B_3 - A_2 B_2 - A_1 B_1 \qquad (8.1.1)$$

In orthogonal coordinate systems, these sums are invariant scalar quantities; a familiar special case is the sum which defines the proper time in special relativity.

$$(ds)^2 = (dt)^2 - (dx)^2 - (dy)^2 - (dz)^2 \qquad (8.1.2)$$

For the more general coordinate systems that we are now considering, which are accelerated, twisted, and stretched, the proper time is defined in terms of products of coordinate displacements and the metric tensor $g_{\mu\nu}$, eq.(7.4.7); we see that the construction of scalar invariants follows a rule more complicated than eq.(8.1.1).

The coordinate displacements are the prototypes of what we will call the contravariant components of a vector. We shall write the indices as superscripts, for example, dx^μ, a conventional notation. What is important is the law of transformation of these contravariant vector components as we change coordinate systems. For the coordinate intervals, this is

$$dx'^\mu = \frac{\partial x'^\mu}{\partial x^\alpha} \, dx^\alpha. \tag{8.1.3}$$

We define a vector function to be a set of four quantities which have the character of coordinate displacements, and transform in the same way as we change coordinates.

$$A^\mu(x') = \frac{\partial x'^\mu}{\partial x^\alpha} \, A^\alpha(x) \tag{8.1.4}$$

We call the quantities A^μ the contravariant components of the vector. We can extend the definitions to tensors of higher rank very easily; for example, a tensor is a function which transforms in the same way as the outer product of two vectors, that is

$$T^{\mu\nu}(x') = \frac{\partial x'^\mu}{\partial x^\alpha} \frac{\partial x'^\nu}{\partial x^\beta} \, T^{\alpha\beta}. \tag{8.1.5}$$

When we compare the transformation law for the metric tensor $g_{\mu\nu}$ with the definition, eq.(8.1.5), we see that $g_{\mu\nu}$ is not the same kind of quantity, since the derivatives appear upside down. However, we have defined a matrix which is the reciprocal of $g_{\mu\nu}$,

$$g^{\nu\alpha} g_{\alpha\beta} = \delta^\nu{}_\beta. \tag{8.1.6}$$

It is not very difficult to prove that this reciprocal is indeed a contravariant tensor, so that it is appropriate to write it with two upper indices, as we have anticipated.

Similarly, it is not difficult to prove that the sums

$$g_{\mu\nu} \, dx^\mu \, dx^\nu = (ds)^2 \tag{8.1.7}$$

and

$$g_{\mu\nu} A^\mu B^\nu$$

are scalar invariants; this comes about because the derivatives appear right side up in one case, and upside down in the other, and they result in Kronecker deltas upon summing.

This suggests that we can use the metric tensor $g_{\mu\nu}$ in order to define a new kind of vector component, having a different law of transformation,

$$(a) \qquad A_\beta = g_{\alpha\beta}A^\alpha,$$

$$(b) \qquad A_\beta(x') = \frac{\partial x^\mu}{\partial x'^\beta}A_\mu(x), \tag{8.1.8}$$

which we shall call the covariant components of the vector. The scalar invariants which may be generated by summing are

$$A_\beta B^\beta. \tag{8.1.9}$$

In the work that follows, it shall be important to keep track of upper and lower indices; in general, only summations over one upper and one lower index will be allowed. For example, in the special case of the orthogonal coordinates of special relativity, the proper time should now be written as

$$(ds)^2 = \eta_{\mu\nu}\,dx^\mu\,dx^\nu. \tag{8.1.10}$$

The tensor $\eta_{\mu\nu}$ is diagonal, and has components $(1, -1, -1, -1)$.

Whenever a vector quantity appears in a physical problem, for example, the vector potential of electrodynamics, it will appear in its definition either as a covariant or contravariant vector. But we can always construct one from the other by use of the metric tensor; we can always lower indices or raise indices at will, by multiplying by either $g_{\mu\nu}$ or its reciprocal. It is possible to construct tensors which are partly covariant, partly contravariant; these are written with some indices up, some indices down, and it is important to write the indices so that there is no question as to their order.

$$g_{\mu\alpha}T^{\mu\nu} = T_\alpha{}^\nu \tag{8.1.11}$$

For the special case of the symmetric tensors $g_{\mu\nu}$ or $g^{\mu\nu}$, we may relax this rule, since raising or lowering one index produces simply the Kronecker delta.

$$g^{\mu\alpha}g_{\alpha\nu} = \delta^\mu{}_\nu = \delta^\mu_\nu \tag{8.1.12}$$

We shall not bother to review the proofs of these relations, since they have been around for many years and are to be found in many books. They were all used by Einstein, who invented the notation that makes it so easy to work with them, and he is a reliable guy when it comes to these things. The placement of the indices, either up or down, is something of a mnemonic, since it corresponds to the placement of the indices in the derivatives that define the transformations, eqs.(8.1.3), (8.1.4), (8.1.5), and (8.1.8).

There is no fundamental physical distinction between the covariant and contravariant components of a vector; they have the same physical

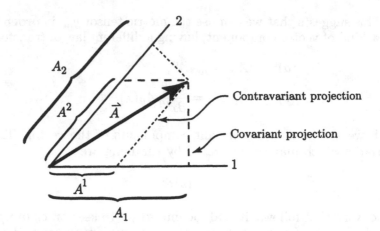

Figure 8.1

content, and it is only the representation which is changed. For the case
of two dimensions, we can easily show graphically how the descriptions
of the vectors differ. Since the transformation is defined as for infinitesimal displacements, we need not worry about the curvature of space; all
that enters is the orthogonality or lack of it. If the coordinate axes do
not intersect at right angles, there are two ways of projecting a physical
displacement onto the axes, either perpendicular to an axis, or parallel
to the others as in Figure 8.1. We see that the tensor components $g_{\mu\nu}$
describe the lack of orthogonality of the coordinates at a given point.

8.2   THE EQUATIONS TO DETERMINE INVARIANTS OF $g_{\mu\nu}$

Now that we have a better understanding of the role of the metric tensor,
we can proceed to study what quantities may be constructed from it
which remain invariant under infinitesimal coordinate transformations.

What we are about to do now is precisely what we did some time ago
in constructing a Lagrangian. Suppose that we make a small change in
the coordinates,

$$x^\mu = x'^\mu + \zeta^\mu(x'), \tag{8.2.1}$$

where ζ^μ is assumed to be small, so we need keep only terms of the first
order in ζ^μ. The derivatives are as follows.

$$\frac{\partial x^\alpha}{\partial x'^\mu} = \delta^\alpha_\mu + \frac{\partial \zeta^\alpha}{\partial x'^\mu} \tag{8.2.2}$$

When we calculate the new $g'_{\mu\nu}$, we get the product of two such derivatives.

$$g'_{\mu\nu}(x') = g_{\alpha\beta}(x' + \zeta) \left(\delta^\alpha_\mu + \frac{\partial \zeta^\alpha}{\partial x'^\mu} \right) \left(\delta^\beta_\nu + \frac{\partial \zeta^\beta}{\partial x'^\nu} \right) \qquad (8.2.3)$$

And if we keep only the terms of zero order and first order in ζ^μ,

$$g'_{\mu\nu}(x') = g_{\mu\nu}(x') + g_{\alpha\nu} \frac{\partial \zeta^\alpha}{\partial x'^\mu} + g_{\mu\beta} \frac{\partial \zeta^\beta}{\partial x'^\nu} + \frac{\partial g_{\mu\nu}}{\partial x'^\sigma} \zeta^\sigma. \qquad (8.2.4)$$

The new $g'_{\mu\nu}$ equals the old $g_{\mu\nu}$, plus some terms of order ζ^μ. As we now ask which functions of $g_{\mu\nu}$ are allowed, if we insist that their form remain invariant, we see that we have arrived at the same mathematical problem that we solved in Lecture 6. The mathematical problem is the same as when we tried finding the Lagrangian which led to a conserved energy-momentum tensor.

There is thus more than one physical point of view which produces the same equation, and which has the same content. We have found that the transformation which arose when we looked for a gravity Lagrangian appears also in the solution of a purely geometrical problem. We guess therefore that some of the physical and geometrical-sounding criteria are equivalent; the self consistency of the previous approach, which made us demand an identically zero divergence, must be equivalent to what we are now doing. What will be the physical significance of the invariants of $g_{\mu\nu}$?

The equations of motion may be derived from the variational principle

$$\delta \int ds = \delta \int \sqrt{g_{\mu\nu}(x) \, dx^\mu dx^\nu} = 0. \qquad (8.2.5)$$

This may be carried out by introducing a parameter u so that the square root of the differential is better defined.

$$\int du \sqrt{g_{\mu\nu}(x) \frac{dx^\mu}{du} \frac{dx^\nu}{du}} \qquad (8.2.6)$$

When the variation is carried out, the following (geodesic) equation is obtained:

$$\frac{d^2 x^\mu}{ds^2} = -\Gamma^\mu_{\sigma\tau} \frac{dx^\sigma}{ds} \frac{dx^\tau}{ds}, \qquad (8.2.7)$$

where

$$\Gamma^\mu_{\sigma\tau} = g^{\mu\nu} [\sigma\tau, \nu].$$

Since the form of this equation remains unchanged as we change the metric tensor in an arbitrary transformation, it must be the invariants of $g_{\mu\nu}$ which contain the physics of the problem.

8.3   ON THE ASSUMPTION THAT SPACE IS TRULY FLAT

Let us try to discuss what it is that we are learning in finding out that these various approaches give the same results. The point of view we had before was that space is describable as the space of Special Relativity; for convenience we shall call it Galilean. In this Galilean space, there might be gravity fields $h_{\mu\nu}$ which have the effect that rulers are changed in length, and clocks go at faster or slower rates. So that in speaking of the results of experiments we are forced to make distinctions between the scales of actual measurements, physical scales, and the scales in which the theory is written, the Galilean scales.

Now, the point is that it is the physical coordinates that must always reproduce the same results. It may be convenient in order to write a theory in the beginning to assume that measurements are made in a space that is in principle Galilean, but after we get through predicting real effects, we see that the Galilean space has no significance.

It makes no sense for us to claim that the choice of coordinates that someone else has made is crazy and cockeyed, just because it does not look Galilean to us. If he insists on treating it as Galilean, and ascribes curvatures to fields, he is perfectly justified also, and it is our space that looks cockeyed to him. For any physical result, one gets the same answer no matter what initial labelling one gives to positions. Therefore, we see that it might be a philosophical improvement if we could state our theory from the beginning in such a way that there is no Galilean space that enters into the specification of the physics; we always deal directly with the physical space of actual measurements.

We may think again of the man who is making measurements with a physical ruler on a hot plate; the ruler obviously changes length as he moves to hotter or cooler regions. But this makes sense to us only because we know of something which can measure distances without being affected by the temperature, namely light. If we, with light measurements, were to inscribe a "truly Euclidean" coordinate system on the plate, the man on the hot plate could quote a temperature field to us, that is, a field which would describe how this ruler changes length as he moves about the plate. If, however, we fool him, and inscribe a distorted coordinate system on the plate, but still tell him it is Euclidean, he will come up with a different temperature field. But there is no way that we can fool him, by inscribing arbitrary coordinates, so that we will ever change the results of physical measurements that he makes entirely by himself. As long as he makes use only of ruler lengths in quoting distances, he will always come up with the same answers, no matter how wacky the temperature fields that he might get by using coordinates that we might specify.

The situation is quite clear in the case of the hot plate, as to what is Euclidean and what is not, only because we have assumed that light measurements are unaffected by heat. In the case of gravity, however, we know of no scale that would be unaffected—there is no "light" unaffected by gravity with which we might define a Galilean coordinate system. Thus, all coordinate systems are equivalent, and they differ only in that different values for the fields are necessary for the description of clock rates or length scales. Once we concentrate on a description of physical measurements, the coordinate system used in the beginning disappears, since it serves only as a convenient labelling, as a bookkeeping device.

There is one case in which there is a significance in a Galilean or a Euclidean system, the limiting case of zero gravity, or the limiting case of uniform temperature on the hot plate. Here, the physical and Euclidean distances follow the same geometry. If we started from a curved labelling of positions, we would find that a certain coordinate transformation enabled us to describe measurements without the use of a field. This is a significant simplification—but again, the simplicity is not due to any inherent validity of a Euclidean description of geometry, but to the fact that it corresponds to a definite physical situation which has a definite physical simplicity.

If the forces are zero everywhere, then the Γ's should be zero everywhere. If the forces are not everywhere zero, there is no possibility of defining the "nicest" coordinate system. It is, however, possible to make them locally zero (principle of equivalence!).

8.4   ON THE RELATIONS BETWEEN DIFFERENT APPROACHES TO GRAVITY THEORY

It is one of the peculiar aspects of the theory of gravitation, that it has both a field interpretation and a geometrical interpretation. Since these are truly two aspects of the same theory we might assume that the Venutian scientists, after developing their completed field theory of gravity, would have eventually discovered the geometrical point of view. We cannot be absolutely sure, since one cannot ever explain inductive reasoning—one cannot ever explain how to proceed, when one knows only a little, in order to learn even more.

In any case, the fact is that a spin-two field has this geometrical interpretation; this is not something readily explainable—it is just marvelous. The geometric interpretation is not really necessary or essential to physics. It might be that the whole coincidence might be understood as representing some kind of gauge invariance. It might be that the relationships between these two points of view about gravity might be transparent after we discuss a third point of view, which has to do with the general properties of field theories under transformations. This point of view will be developed more fully later—we discuss it here so as to get a feeling for some directions which we might take in attempting to understand how gravity can be both geometry and a field.

Let us consider just what a gauge invariance is. As usually stated in electrodynamics, it means that if we replace a vector potential A by

$$A' = A + \nabla X, \tag{8.4.1}$$

the equations of the field and physical effects remain unchanged, in terms of the new vector potential A'. This may be related to the property of phase invariance of amplitudes. Let us now see what happens to quantum-mechanical amplitudes; it is quite certain that if we use

$$\psi' = e^{ia}\psi$$

in calculating a probability, nothing changes in the predicted physics. In general, a constant a makes no difference. What happens if instead of a constant a, we use a function X which varies from point to point in space? The equations always involve gradients of ψ' which are

$$\nabla\psi' = e^{iX}\left(\nabla\psi + i\psi\nabla X\right). \tag{8.4.2}$$

However, the operator $(\nabla - iA')$ leaves something which has changed only by a phase,

$$(\nabla - iA')\psi' = e^{iX}(\nabla - iA)\psi. \tag{8.4.3}$$

So that, if there exists a vector field which couples as we have supposed, the equations are invariant under space-time dependent phase transformations of the ψ fields.

The Yang-Mills vector meson theory is an attempt to extend the idea of a transformation gauge by considering in a similar way the invariance of nuclear interactions under isotopic spin. If an amplitude for a proton is represented by ψ, then

$$\psi' = e^{i\vec{\tau}\cdot\vec{a}}\psi \tag{8.4.4}$$

describes an object which is partly a proton, and partly a neutron. If $\vec{a}$ is a constant vector in isospin space, the invariance of the nuclear forces with respect to changes in isotopic spin means that the new object ψ' acts in all nuclear respects like ψ. The proposal of Yang and Mills is that a field should be added to the Lagrangian, in such a way that a *space-dependent* phase change $(\vec{a} \to \vec{X})$ makes no difference in the equations.

How can such ideas be connected to gravity? The equations of physics are invariant when we make coordinate displacements any constant amount a^{μ}.

$$x'^{\mu} = x^{\mu} + a^{\mu} \tag{8.4.5}$$

To make it formally more like the phase and isospin transformations, one might use the momentum representation so that the translation operator is

$$\exp\left(ip_{\mu}a^{\mu}\right).$$

On the other hand, it is possible to investigate how we might make the equations of physics invariant when we allow space dependent *variable* displacements $(a^\mu \rightarrow \zeta^\mu)$. The search will be for a more complete Lagrangian; the new terms that are needed are precisely those of a gravity field. Thus, gravity is that field which corresponds to a gauge invariance with respect to displacement transformations.

8.5   THE CURVATURES AS REFERRED TO TANGENT SPACES

We can argue as Einstein, geometrically, and ask for curvatures and such things, in terms of limiting cases of radii and circumference. Just to show that the thing is not so complicated, we will do it. Now that we know what we are doing, we can make general coordinate transformations. We have talked about how many derivatives there were. We have been able to say that we can set

$$g^{\text{o}}_{\mu\nu} = \eta_{\mu\nu} \qquad\qquad (8.5.1)$$

by adjusting the sixteen derivatives $(\partial x^\alpha / \partial x'^\nu)$. We also assume that we can adjust the forty second derivatives $(\partial^2 x^\alpha / \partial x'^\mu \partial x'^\nu)$, to make all first derivatives of $g^{\text{o}}_{\mu\nu}$ zero. Then, there are eighty adjustable third derivatives, and one hundred derivatives of the $g_{\mu\nu}$'s. It is twenty linear combinations of these second derivatives which are the things that may have a geometrical definition. They cannot be reduced away by transformations of coordinates. What we are looking for is an expression for twenty such quantities in terms of the initial $g_{\mu\nu}$. We do this in three steps, going backwards, so to speak. First we assume that we have adjusted the first and second derivatives (at a point by transforming to Riemann normal coordinates) so that $g^{\text{o}}_{\mu\nu} = \eta_{\mu\nu}$ and $g^{\text{o}}_{\mu\nu,\sigma} = 0$, and find the expression for twenty quantities. Then we will worry about precisely how we have succeeded in making these adjustments, from the arbitrary initial coordinates, to get the twenty quantities in terms of the original $g_{\mu\nu}$.

First we discuss the geometrically definable quantities in terms of coordinates in the tangent space at a point. What we are doing in four dimensions is analogous to the following situation in two dimensions. The curved space is a surface, and we compare the geometry of the surface of the point to the geometry as viewed from the tangent plane as in Figure 8.2. The initial coordinates in the curved surface are in general not orthogonal and not properly oriented to allow the simplest description of the geometry in terms of the invariant $1/(R_1 R_2)$. The first step is a determination of this intrinsic curvature in terms of the initial geometry.

Only the second derivatives begin to describe the deviation from flatness of the curved space at the point. The curvatures are precisely a measure of the local mismatch between the surface and the tangent plane. They are a description of the essential character of the space at the point.

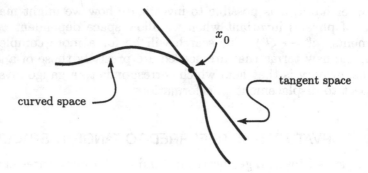

x_0

tangent space

curved space

Figure 8.2

Since only first, second, and third derivatives of the coordinates enter, we have a sufficiently general transformation in

$$x^\alpha = x'^\alpha + \frac{1}{2}a^\alpha{}_{\mu\nu}x'^\mu x'^\nu + \frac{1}{6}b^\alpha{}_{\mu\nu\sigma}x'^\mu x'^\nu x'^\sigma. \qquad (8.5.2)$$

It is our task to pick out the twenty significant combinations. The first derivatives are

$$\frac{\partial x^\alpha}{\partial x'^\mu} = \delta^\alpha{}_\mu + a^\alpha{}_{\mu\nu}x'^\nu + \frac{1}{2}b^\alpha{}_{\mu\nu\sigma}x'^\nu x'^\sigma. \qquad (8.5.3)$$

In this particular tangent space, the metric tensors can be sufficiently generally written as

$$\begin{aligned}
g'_{\alpha\beta} &= \eta_{\alpha\beta} + \frac{1}{2}x'^\sigma x'^\tau g^{o\prime}{}_{\alpha\beta,\sigma\tau}, \\
g_{\alpha\beta} &= \eta_{\alpha\beta} + \frac{1}{2}x^\sigma x^\tau g^{o}{}_{\alpha\beta,\sigma\tau}.
\end{aligned} \qquad (8.5.4)$$

A superscript o denotes the particular quantity evaluated at the point of tangency x_0. We get the proper invariant combinations by considering that we have two arbitrary coordinate systems in the tangent space, and demand that the same formulae should work in both cases. Since the spaces are tangent, the coordinates must differ only quadratically, so that

$$a^\alpha{}_{\mu\nu} = 0. \qquad (8.5.5)$$

Then, it is only necessary to put one transformation inside the other. Substituting the law of transformation of the $g_{\mu\nu}$'s, using the expressions, eq.(8.5.3), for the derivatives, we find that

$$g'_{\alpha\beta} = g_{\mu\nu}\frac{\partial x^\mu}{\partial x'^\alpha}\frac{\partial x^\nu}{\partial x'^\beta} = \eta_{\alpha\beta} + \frac{1}{2}\left(g^o_{\alpha\beta,\sigma\tau} + \eta_{\alpha\nu}b^\nu_{\beta\sigma\tau} + \eta_{\beta\nu}b^\nu_{\alpha\sigma\tau}\right)x'^\sigma x'^\tau.$$
$$(8.5.6)$$

The second derivative of the g's compares as follows:

$$g'^o_{\alpha\beta,\sigma\tau} = g^o_{\alpha\beta,\sigma\tau} + b_{\alpha\beta\sigma\tau} + b_{\beta\alpha\sigma\tau}, \qquad (8.5.7)$$

where

$$b_{\alpha\beta\sigma\tau} \equiv \eta_{\alpha\nu}\, b^\nu{}_{\beta\sigma\tau}.$$

Now that we have this expression, we want linear combinations of the g's which do not have the b's. We make use of the fact that the b's are completely symmetric in their last three indices, while the $g_{\alpha\beta,\tau}$ is symmetric only in $\sigma\tau$. Try flipping indices $(\beta \leftrightarrow \sigma)$ and subtracting terms

$$g'^o_{\alpha\beta,\sigma\tau} - g'^o_{\alpha\sigma,\beta\tau} - g^o_{\alpha\beta,\sigma\tau} + g^o_{\alpha\sigma,\beta\tau} = b_{\beta\alpha\sigma\tau} - b_{\sigma\alpha\beta\tau}. \qquad (8.5.8)$$

The indices $(\alpha\tau)$ are absolutely symmetric in the right-hand side of this expression, but not necessarily on the left. Therefore, in antisymmetrizing $(\alpha - \tau)$ we get the following expression:

$$R_{\alpha\tau\beta\sigma} = \frac{1}{2}\left(g^o_{\alpha\beta,\sigma\tau} - g^o_{\alpha\sigma,\beta\tau} - g^o_{\tau\beta,\sigma\alpha} + g^o_{\tau\sigma,\beta\alpha}\right), \qquad (8.5.9)$$

a thing which equals itself in the primed system. These are the twenty linear combinations which we were looking for. This quantity is not a tensor; it is not general enough; it is defined only in a place having zero net fields. These are in a sense the irreducible pieces of the gravitational tensor, those which can't be removed by a transformation. They represent the purely tidal forces. Thus, we now have a definite prescription to find the curvatures. First, find the transformation (to Riemann normal coordinates) that makes $g_{\mu\nu}$'s into $\eta_{\mu\nu}$'s and first derivatives zero. Then, in terms of the transformed $g_{\mu\nu}$'s, the curvatures are given by eq.(8.5.9). These are the same in any coordinate system. The remaining problem is to express the $R_{\alpha\tau\beta\sigma}$ in terms of the original arbitrary coordinates, and the original $g_{\mu\nu}$'s.

8.6   THE CURVATURES REFERRED TO ARBITRARY COORDINATES

The deduction of expressions for the curvatures in terms of general coordinates proceeds most smoothly in a stepwise fashion. Next, we remove the restriction on the first derivatives (they may now be nonzero), but keep the coordinates orthogonal locally; the expressions of $g_{\mu\nu}$ and $g'_{\mu\nu}$ are

$$g_{\alpha\beta} = \eta_{\alpha\beta} + g^{\circ}{}_{\alpha\beta,\mu}x^{\mu} + \frac{1}{2}g^{\circ}{}_{\alpha\beta,\mu\nu}x^{\mu}x^{\nu},$$

$$g'_{\alpha\beta} = \eta_{\alpha\beta} + \frac{1}{2}g'^{\circ}{}_{\alpha\beta,\sigma\tau}x'^{\sigma}x'^{\tau}. \tag{8.6.1}$$

The g' is the same as before, with the zero first derivatives. Since we have already adjusted the second derivatives, we need now only consider the transformations of the type

$$x^{\alpha} = x'^{\alpha} + \frac{1}{2}a^{\alpha}{}_{\mu\nu}x'^{\mu}x'^{\nu}. \tag{8.6.2}$$

The cubic terms will not affect the right. The expression for the first derivatives,

$$\frac{\partial x^{\alpha}}{\partial x'^{\mu}} = \delta^{\alpha}{}_{\mu} + a^{\alpha}{}_{\mu\nu}x'^{\nu}, \tag{8.6.3}$$

is now plugged into the equation expressing g' in terms of g.

$$\eta_{\alpha\beta} + \frac{1}{2}g'^{\circ}{}_{\alpha\beta,\sigma\tau}x'^{\sigma}x'^{\tau} = \eta_{\alpha\beta} + \left(g^{\circ}_{\alpha\beta,\sigma} + a_{\beta\alpha\sigma} + a_{\alpha\beta\sigma}\right)x'^{\sigma}$$
$$+ x'^{\sigma}x'^{\tau}\left[a^{\rho}{}_{\alpha\sigma}a_{\rho\beta\tau} + a^{\rho}{}_{\alpha\tau}g^{\circ}_{\rho\beta,\sigma} + a^{\rho}{}_{\beta\tau}g^{\circ}_{\rho\alpha,\sigma} + \frac{1}{2}a^{\rho}{}_{\sigma\tau}g^{\circ}_{\alpha\beta,\rho} + \frac{1}{2}g_{\alpha\beta,\sigma\tau}\right], \tag{8.6.4}$$

where $a_{\alpha\beta\sigma} = \eta_{\alpha\mu}a^{\mu}{}_{\beta\sigma}$. The first derivative terms will become zero with the following choice of $a^{\alpha}{}_{\mu\nu}$:

$$a_{\beta\alpha\sigma} + a_{\alpha\beta\sigma} = -g^{\circ}_{\alpha\beta,\sigma}. \tag{8.6.5}$$

We need to solve this equation so that the $a^{\alpha}_{\mu\nu}$ are expressed in terms of the $g^{\circ}_{\alpha\beta,\sigma}$ of the initial coordinate system. This is done by the usual tricks; subtract the equation obtained by reversing (α, σ), then collect terms, etc., and we obtain the solution

$$a_{\sigma\alpha\beta} = -\frac{1}{2}\left[g^{\circ}_{\sigma\alpha,\beta} + g^{\circ}_{\sigma\beta,\alpha} - g^{\circ}_{\alpha\beta,\sigma}\right] = -[\alpha\beta,\sigma]^{\circ}. \tag{8.6.6}$$

Eq.(8.6.4) now tells us that $g'^{\circ}{}_{\alpha\beta,\sigma\tau}$ is (twice) the expression in brackets [ ] on the right side of eq.(8.6.4) with eq.(8.6.6) substituted for $a_{\sigma\alpha\beta}$. This

$g'^0_{\alpha\beta,\sigma\tau}$ may now be substituted for the $g^0_{\alpha\beta,\sigma\tau}$ in eq.(8.5.9) to find for the curvatures, in terms of the old coordinates (restricted only in that they must be locally orthogonal), the following

$$R_{\alpha\tau\beta\sigma} = \frac{1}{2}\left(g_{\alpha\beta,\sigma\tau} - g_{\alpha\sigma,\beta\tau} - g_{\tau\beta,\alpha\sigma} + g_{\tau\sigma,\alpha\beta}\right)$$

$$+ [\rho\sigma,\alpha]\,\eta^{\rho\lambda}\,[\tau\beta,\lambda] - [\rho\beta,\alpha]\,\eta^{\rho\lambda}\,[\tau\sigma,\lambda]. \quad (8.6.7)$$

All that remains is orthgonalizing the initially general coordinates. This can be done by a linear transformation:

$$x^\alpha = L^\alpha{}_\mu x'^\mu. \tag{8.6.8}$$

All that needs to be done is choose

$$g'^0_{\alpha\beta} = \eta_{\alpha\beta}, \tag{8.6.9}$$

and rewrite everything else. The derivatives are simply the L's, as follows:

$$g'_{\mu\nu} = L^\alpha{}_\mu L^\beta{}_\nu g_{\alpha\beta} = \eta_{\mu\nu}. \tag{8.6.10}$$

Among the things that may be deduced is that

$$\eta^{\alpha\beta} L_\alpha{}^\sigma L_\beta{}^\mu = g^{\sigma\mu}. \tag{8.6.11}$$

What happens to the various terms?

$$\frac{\partial}{\partial x'^\alpha} = \frac{\partial}{\partial x^\mu}\frac{\partial x^\mu}{\partial x'^\alpha} = L^\mu{}_\alpha \frac{\partial}{\partial x^\mu} \tag{8.6.12}$$

Therefore, (here, Latin indices represent indices of primed coordinates)

$$g'_{mn,st} = L^\sigma{}_s L^\tau{}_t L^\mu{}_m L^\nu{}_n g_{\mu\nu,\sigma\tau} \tag{8.6.13}$$

$$a'_{rmn} = L^\rho{}_r L^\mu{}_m L^\nu{}_n a_{\rho\mu\nu} \tag{8.6.14}$$

$$\eta^{rq} a'_{rmn} a'_{qst} = \eta^{rq} L^\rho{}_r L^\lambda{}_q L^\mu{}_m L^\nu{}_n L^\sigma{}_s L^\tau{}_t a_{\rho\mu\nu} a_{\lambda\sigma\tau} \tag{8.6.15}$$

When we plug this into the R's, we find that R is no longer an invariant. The final expression for R is (using eq.(8.6.11))

$$R_{\alpha\tau\beta\sigma} = \frac{1}{2}\left(g_{\alpha\beta,\sigma\tau} - g_{\alpha\sigma,\beta\tau} - g_{\tau\beta,\alpha\sigma} + g_{\tau\sigma,\alpha\beta}\right)$$

$$+ [\rho\sigma,\alpha]g^{\rho\lambda}[\tau\beta,\lambda] - [\rho\beta,\alpha]g^{\rho\lambda}[\tau\sigma,\lambda], \quad (8.6.16)$$

and the transformation law is

$$R'_{mnst} = L^\mu{}_m L^\nu{}_n L^\sigma{}_s L^\tau{}_t R_{\mu\nu\sigma\tau}. \tag{8.6.17}$$

8.7   PROPERTIES OF THE GRAND CURVATURE TENSOR

Although the quantities $R_{\mu\nu\sigma\tau}$ are not invariants, they are a tensor, as may be judged from the transformation law, eq.(8.6.17). It may easily be shown that it does contain only twenty quantities, as we have previously claimed. The expressions eq.(8.5.9) have been obtained by antisymmetrizing in (α, τ) and subsequently in (β, σ). The following symmetries follow.

$$
\begin{aligned}
R_{\alpha\tau\beta\sigma} &= -R_{\tau\alpha\beta\sigma} &\text{(a)} \\
&= -R_{\alpha\tau\sigma\beta} &\text{(b)} \\
&= +R_{\beta\sigma\alpha\tau} &\text{(c)}
\end{aligned}
\tag{8.7.1}
$$

The following algebraic relation is also implicit in eq.(8.5.9) (and therefore in eq.(8.6.16)):

$$
R_{\alpha\tau\beta\sigma} + R_{\alpha\sigma\tau\beta} + R_{\alpha\beta\sigma\tau} = 0.
\tag{8.7.2}
$$

Let us now count the number of independent components. The first index may not equal the second, and the third may not equal the fourth. Only the antisymmetric combinations are nonzero—we recall there are six possible nonzero components for an antisymmetric tensor of second rank, so that except for the symmetry for exchange of the first pair with the second pair, there would be 36 possibilities; this last symmetry, eq.(8.7.1c), reduces the number to $(6 \times 7)/2 = 21$. The algebraic relation, eq.(8.7.2), contains only one nontrivial restriction. If two indices are the same, eq.(8.7.2) is an identity because of the symmetries in eq.(8.7.1). For example,

$$
R_{1\tau1\sigma} + R_{1\sigma\tau1} + R_{11\sigma\tau} = R_{1\tau1\sigma} - R_{1\tau1\sigma} + 0 = 0.
\tag{8.7.3}
$$

So that all indices must be different for the algebraic relation to be meaningful. But when all indices are different, (1,2,3,4), there is only one extra equation. So that in general there are only twenty independent components of the Grand Curvature Tensor (Riemann Tensor).

What we want for the construction of our theory, is not a tensor, but a completely invariant thing to put into the Lagrangian. (Instead, Einstein said that the Stress-Energy Tensor equals another tensor, which is derivable from the curvature tensor.) The least-action principle must involve an integral over all space, which must be completely invariant to transformations. The integrand must be a world scalar.

$$
\int dx\, dy\, dz\, dt\ (\text{Scalar}) = (\text{Scalar Invariant})
\tag{8.7.4}
$$

We get such a scalar by raising the indices of the curvature tensor and contracting. We may for example raise the first index,

$$
g^{\alpha\lambda} R_{\alpha\tau\beta\sigma} = R^{\lambda}{}_{\tau\beta\sigma}.
\tag{8.7.5}
$$

But when we contract at this point, in the first pair of indices the thing unfortunately vanishes.

$$R^{\tau}{}_{\tau\beta\sigma} \equiv 0 \qquad (8.7.6)$$

What is done is to first reduce the rank of the tensor and contract in the first and last index.

$$g^{\alpha\sigma} R_{\alpha\tau\beta\sigma} = R_{\tau\beta} \qquad (8.7.7)$$

(Note that the same letter R is conventionally used for all of these tensors deduced from the curvature tensor.) This second rank tensor (the Ricci tensor) is symmetric. Then we reduce the rank again to obtain our scalar (the "scalar curvature") for the integrand,

$$g^{\alpha\sigma} g^{\tau\beta} R_{\alpha\tau\beta\sigma} = g^{\tau\beta} R^{\sigma}{}_{\tau\beta\sigma} = R^{\sigma\beta}{}_{\beta\sigma} = R. \qquad (8.7.8)$$

Now, the volume integral of this scalar is not an invariant, because the volume element is not a scalar; $dx\,dy\,dz\,dt$ changes as we change coordinates by the determinant of the $L_{\alpha}{}^{\mu}$. So that the invariant integral is

$$\int dx\,dy\,dz\,dt\,R\,\sqrt{-g}. \qquad (8.7.9)$$

This is the Einstein-Hilbert action for empty space [Hilb 15].

9.1 MODIFICATIONS OF ELECTRODYNAMICS REQUIRED BY THE PRINCIPLE OF EQUIVALENCE

The Principle of Equivalence postulates that an acceleration shall be indistinguishable from gravity by any experiment whatsoever. In particular, it cannot be distinguished by observing electromagnetic radiation. There is evidently some trouble here, since we have inherited a prejudice that an accelerating charge should radiate, whereas we do not expect a charge lying in a gravitational field to radiate. This is, however, not due to a mistake in our statement of equivalence but to the fact that the rule of the power radiated by an accelerating charge,

$$\frac{dW}{dt} = \frac{2}{3}\frac{e^2}{c^3}a^2,$$ (9.1.1)

has led us astray. This is usually derived from calculating the flow from Poynting's theorem far away, and it is only valid for cyclic motions, or at least motions which do no grow forever in time (as a constant acceleration does). It does not suffice to tell us "when" the energy is radiated. This can only be determined by finding the force of radiation resistance, which is $2/3\ e^2/c^3\ \dot{a}$. For it is work against this force which represents energy

123

loss. For constant acceleration this force is zero. Generally the work done against it can be written

$$\frac{dW}{dt} = -\frac{2}{3}\frac{e^2}{c^3}\,\vec{v}\cdot\dot{\vec{a}} = \frac{2}{3}\frac{e^2}{c^3}\,\dot{\vec{a}}\cdot\dot{\vec{a}} - \frac{2}{3}\frac{e^2}{c^3}\frac{d}{dt}(\vec{v}\cdot\dot{\vec{a}}),\qquad(9.1.2)$$

giving a correct expression for dW/dt. For cyclic or limited motions, the average contribution of the last term over the long run is small or zero (over one cycle since $\vec{v}\cdot\dot{\vec{a}}$ is restored to its original value, its contribution vanishes) and the simpler eq.(9.1.1) suffices.

Of course, in a gravitational field the electrodynamic laws of Maxwell need to be modified, just as ordinary mechanics needed to be modified to satisfy the principle of relativity. After all, the Maxwell equations predict that light should travel in a straight line—and it is found to fall towards a star. Clearly, some interaction between gravity and electrodynamics must be included in a better statement of the laws of electricity, to make them consistent with the principle of equivalence.

We shall not have completed our theory of gravitation until we have discussed these modifications of electrodynamics, and also the mechanisms of emission, reception, and absorption of gravitational waves.

9.2   COVARIANT DERIVATIVES OF TENSORS

In the previous lecture, we have seen how the notion of the curvature of space arose in discussing geometrical measurements. We can get a better idea of the way in which four-dimensional curvatures will affect our viewpoint on physics by considering a more conventional approach, which is to define curvatures by what happens to a vector as we move about in the space. Let us imagine again our two-dimensional world. If we use flat Euclidean coordinates, a constant vector field existing throughout the space is described by constant components. If we use some other coordinates, curved in general, the constant vector field is described by components which vary from point to point. A familiar example is that of plane polar coordinates, in which a constant vector F is described by the components,

$$(A\cos\theta + B\sin\theta) = F_r,$$
$$(B\cos\theta - A\sin\theta) = F_\theta.\qquad(9.2.1)$$

The first thing we must do is to develop formulae which allow us to compare the physically meaningful difference between a tensor at one point and its value at neighboring points, that is, we want to describe the variation of the tensor in a way which cancels out the changes in components induced by arbitrary choices of coordinates; for example we want to compare a vector at x^μ to another vector at a point an infinitesimal displacement dx^μ away, by carrying one of the vectors, keeping it constant (more precisely—keeping it parallel to itself) to the other point.

For a scalar function, a tensor of zero rank, there is no problem. The ordinary gradient transforms as

$$\frac{\partial\phi}{\partial x'^\nu} = \frac{\partial x^\sigma}{\partial x'^\nu}\frac{\partial\phi}{\partial x^\sigma}, \tag{9.2.2}$$

so that the gradient of a scalar is evidently a covariant vector. However, the ordinary gradients of vectors or higher-rank tensor quantities are not tensors; they have a transformation law that contains terms depending on the accidents of the coordinates. We deduce a proper form for the derivative by considering how things look in the tangent space. Since the tangent space is flat, the derivatives of the components contain no terms due to the curving of the coordinates, and the gradients of vectors with respect to flat coordinates are tensors. We get a formula for these tensors in any coordinates by transforming back from the flat space to general coordinates. As usual, we employ expansions to do this. (The primed coordinates are in the flat space.) Let

$$x^\nu = x'^\nu + \frac{1}{2}a^\nu{}_{\sigma\tau}x'^\sigma x'^\tau + \cdots,$$
$$\frac{\partial x^\nu}{\partial x'^\mu} = \delta^\nu_\mu + a^\nu{}_{\sigma\mu}x'^\sigma + \cdots. \tag{9.2.3}$$

Working near the origin of the expansion,

$$A_\mu(x) = \frac{\partial x^\nu}{\partial x'^\mu}A'_\nu(x').$$

And, because we can write the derivative by use of eq.(9.2.3),

$$A_\mu(x) = A'_\mu(x') + a^\nu_{\mu\lambda}x'^\lambda A'_\nu(x') + \cdots. \tag{9.2.4}$$

We now take the gradient of this expression with respect to the general coordinate and evaluate it at the origin.

$$\left(\frac{\partial A_\mu}{\partial x^\sigma}\right)_o = \frac{\partial}{\partial x'^\tau}\left(A'_\mu(x') + a^\nu_{\mu\lambda}x'^\lambda A'_\nu(x')\right)_o\left(\frac{\partial x'^\tau}{\partial x^\sigma}\right)_o$$
$$= \frac{\partial A'_\mu(x')}{\partial x'^\sigma} + a^\nu_{\mu\sigma}A'_\nu(x') \tag{9.2.5}$$

It is because this quantity is evaluated at the origin, that all terms linear in x' are zero. Thus, we obtain a formula for the "flat-space" derivative in terms of the general coordinates.

$$\left[\frac{\partial A_\mu}{\partial x^\sigma} - a^\nu_{\mu\sigma} A_\nu\right] = \frac{\partial A'_\mu}{\partial x'^\sigma} \tag{9.2.6}$$

If we now write $a^\nu_{\mu\sigma}$ in terms of the metric tensors, we obtain a formula for the truer derivative. This quantity is a tensor, and it is known as the covariant derivative of the vector A_τ. To distinguish it from the gradients, we shall use a semicolon to denote covariant differentiation.

$$A_{\mu;\tau} \equiv \frac{\partial A_\mu}{\partial x^\tau} - \Gamma^\sigma_{\mu\tau} A_\sigma \tag{9.2.7}$$

The demonstration that the quantity is a tensor is tedious but straight-forward; all that is required is to convert all coordinates to flat space, and to actually compute the derivative, then to check upon the law of transformation. The rule for differentiating a contravariant vector is similar.

$$A^\mu_{;\sigma} = \frac{\partial A^\mu}{\partial x^\sigma} + \Gamma^\mu_{\sigma\tau} A^\tau \tag{9.2.8}$$

It may be proved most easily by starting from eq.(9.2.7) and using the metric tensor to raise and lower indices; the rearrangement of the metric tensors result in a Γ reversed in sign. To compute the covariant derivative of a tensor having many indices, the rule is

$$T^{\mu\nu}_{\ \ \rho;\lambda} = \frac{\partial T^{\mu\nu}_{\ \ \rho}}{\partial x^\lambda} + \Gamma^\mu_{\lambda\sigma} T^{\sigma\nu}_{\ \ \rho} + \Gamma^\nu_{\lambda\sigma} T^{\mu\sigma}_{\ \ \rho} - \Gamma^\sigma_{\lambda\rho} T^{\mu\nu}_{\ \ \sigma}. \tag{9.2.9}$$

In other words, each index brings an added term involving a Γ and the tensor itself. There is hardly any mnemonic necessary; indices can be up or down only in one way and it is only the identification + with up, − with down that needs to be memorized.

The most familiar example of such transformations is the formula for the curl of a vector in spherical coordinates; these formulae always involve ordinary derivatives multiplied by the values of the components of the vector.

One more relationship of covariant derivatives is useful. Since the covariant derivatives of the metric tensors are zero, as may be easily shown,

$$g^{\mu\nu}_{\ \ ;\sigma} = 0, \tag{9.2.10}$$

the following rule applies to the covariant derivative of a product.

$$(A^\mu B^\nu)_{;\sigma} = A^\mu B^\nu_{\ ;\sigma} + A^\mu_{\ ;\sigma} B^\nu \tag{9.2.11}$$

To show such identities relating tensor quantities, it is always permissible to choose a coordinate system that makes the proof easier; tensors are mathematical quantities so constructed that a tensor identity proved in one coordinate system is true in all. The last equation is easily proved by going to the flat tangent space; the covariant derivative is equal to the ordinary derivative in this space.

One effect of the curvatures is that a second covariant differentiation does not commute with the first. We may explicitly compute what such things are in general by repeated use of eq.(9.2.9). The first stage gives us

$$
\begin{aligned}
A^\mu{}_{;\sigma;\tau} &= [A^\mu{}_{;\sigma}]_{;\tau} \\
&= \frac{\partial[A^\mu{}_{;\sigma}]}{\partial x^\tau} + \Gamma^\mu_{\tau\lambda}[A^\lambda{}_{;\sigma}] - \Gamma^\lambda_{\sigma\tau}[A^\mu{}_{;\lambda}],
\end{aligned}
\tag{9.2.12}
$$

and the second differentiation gives us

$$
\begin{aligned}
A^\mu{}_{;\sigma;\tau} &= \frac{\partial^2 A^\mu}{\partial x^\tau \partial x^\sigma} + \frac{\partial}{\partial x^\tau}\left(\Gamma^\mu_{\sigma\lambda}A^\lambda\right) \\
&\quad + \Gamma^\mu_{\tau\lambda}\left(\frac{\partial A^\lambda}{\partial x^\sigma} + \Gamma^\lambda_{\sigma\rho}A^\rho\right) - \Gamma^\lambda_{\sigma\tau}\left(\frac{\partial A^\mu}{\partial x^\lambda} + \Gamma^\mu_{\lambda\rho}A^\rho\right).
\end{aligned}
\tag{9.2.13}
$$

The noncommutation of the order of the covariant derivatives is seen when we compute the difference,

$$
A^\mu{}_{;\sigma\tau} - A^\mu{}_{;\tau\sigma} = \left[\Gamma^\mu_{\sigma\rho,\tau} - \Gamma^\mu_{\tau\rho,\sigma} + \Gamma^\mu_{\tau\lambda}\Gamma^\lambda_{\rho\sigma} - \Gamma^\mu_{\sigma\lambda}\Gamma^\lambda_{\rho\tau}\right] A^\rho.
\tag{9.2.14}
$$

The factor multiplying A^ρ must be a tensor, since the quantity on the left is a difference of tensors. It is precisely the curvature tensor, so that

$$
A^\mu{}_{;\sigma\tau} - A^\mu{}_{;\tau\sigma} = R^\mu{}_{\rho\sigma\tau}A^\rho.
\tag{9.2.15}
$$

9.3   PARALLEL DISPLACEMENT OF A VECTOR

The fact that the curvature tensor appears as connecting the second covariant derivatives serves as a clue that enables us to give another useful geometrical picture of curvature. The noncommuting property of the second derivatives represents a limit of differences in the vector as we move first a displacement along axis σ and then along τ, or first along τ and then along σ. If the coordinates are flat, for a constant vector there is no difference. If we have curved space as we take these displacements in different orders, we find a different resulting vector. The importance of these considerations in making physical statements becomes apparent when we realize that we have no physical way of defining a "truly constant" vector

Figure 9.1

field except as a vector field whose components have zero derivatives in the tangent space.

How a curvature appears in considering displacements of a vector keeping it parallel to itself as one moves about the surface, is well illustrated in a spherical geometry. We shall imagine that we carry a little vector from the North pole down to the equator, then along the equator through an angle θ, and back up to the North pole as in Figure 9.1, always translating it so it keeps parallel, which in this case means always pointing straight South. When we arrive back at the North pole, we see that our vector has turned through an angle θ. The curvature K of a surface is defined in terms of the angle through which a vector is turned as we take it around an infinitesimal closed path. For a surface,

$$\delta\theta = (\text{Area Enclosed}) \cdot K. \tag{9.3.1}$$

For the case of a triangle on a spherical surface, this angle is precisely the excess (over 180°) of the sum of angles of the triangle. For a spherical surface, the curvature K simply equals $1/R^2$.

The generalized definition of curvature of a many-dimensional surface will be given in terms of the change in a vector as it is carried about a closed path, keeping it parallel to itself. Since the orientation of a path lying on a definite plane depends on two coordinate axes, we see that the curvature will have the general character of a fourth rank tensor. In a three-dimensional space, we might lay out a spherical surface by going "radially" outward from a point for a given measured distance along the shortest measured paths (geodesics). The components of curvature along different directions would correspond to the fractional deviation from 2π of the lengths of the great circles of the spherical surface.

Visualization in terms of a simpler space embedded in a higher-dimensional space requires one extra dimension for each independent

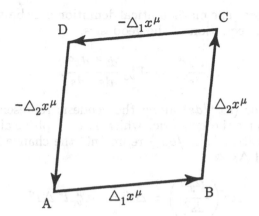

Figure 9.2

component of the metric tensor. For two-dimensional spaces, there are three components of the metric, and hence three dimensions are sufficient. For three dimensions, the metric tensor has six independent components, and for four dimensions, there are ten independent metric components.

The definition of curvature components in terms of the change in a vector in going around a path is more general than the definition in terms of defects in circumferences, which doesn't reproduce all features of the curvature.

The connection with second covariant derivatives may be easily computed as we consider successive displacements of a vector, keeping it parallel to itself. As we go about the path in Figure 9.2, the difference in the vector after having gone around must be

$$\delta^2 A^\mu = R^\mu{}_{\nu\sigma\tau} A^\nu \triangle_1 x^\sigma \triangle_2 x^\tau. \qquad (9.3.2)$$

Since the curvature is a tensor antisymmetric in (σ, τ), the bilinear products $\triangle_1 x^\sigma \triangle_2 x^\tau$ can be replaced by $1/2(\triangle_1 x^\sigma \triangle_2 x^\tau - \triangle_1 x^\tau \triangle_2 x^\sigma)$, which are one half of the components of the area of the parallelogram. The indices of the tensor have a significance which is not difficult to describe verbally; in taking a vector about a small loop in the $(\sigma\tau)$ plane, the μ component of the vector changes by an amount proportional to a sum over the other components A^ν, $R^\mu{}_{\nu\tau\sigma} A^\nu$, and to the area of the loop.

We have talked a great deal about displacing a vector parallel to itself without making the concept mathematically definite. In more intuitive terms, it simply means that we carry the tip of the arrow and the back of the arrow an equal displacement as nearly as we can, along a straight line, that is, the geodesic. The mathematical definition may best be understood by considering the equation of the geodesics,

$$\frac{d^2 x^\mu}{ds^2} = -\Gamma^\mu_{\nu\sigma} \frac{dx^\nu}{ds} \frac{dx^\sigma}{ds}. \tag{9.3.3}$$

Clearly, the vector (dx^μ/ds) along the geodesic represents a tangential velocity, Δt^μ, along the geodesic, which is the "physical" straight line. The second derivative $(d^2 x^\mu/ds^2)$ represents the change in this velocity in a time interval Δs.

$$\Delta s \left(\frac{d^2 x^\mu}{ds^2} \right) = \Delta t^\mu = -\Gamma^\mu_{\nu\sigma} t^\nu \Delta x^\sigma \tag{9.3.4}$$

The change is proportional to the vector itself, t^ν, and the displacements Δx^σ. The definition of a parallel displacement is analogous; we say the vector A'^μ is the result of displacing A^μ parallel to itself if

$$A'^\mu = A^\mu + \delta A^\mu$$

where
$$\delta A^\mu = -\Gamma^\mu_{\sigma\nu} A^\sigma \Delta x^\nu. \tag{9.3.5}$$

It may easily be shown that as we move a set of vectors about a closed loop, displacing each parallel to itself, the relations between the vectors do not change, so that the whole space defined by a set of vectors rotates as we go about the loop; this is the entire change induced by the displacements. The proof is done by showing that all invariant scalars,

$$B^\mu A^\nu g_{\mu\nu}, \tag{9.3.6}$$

remain unchanged. This means that the lengths of vectors, and the angles between vectors, are preserved. The only change which is allowed is like a rigid rotation of the whole space.

It is possible that the topological properties of the space are not completely defined by the local curvature. For example, we have deduced that vector lengths are preserved and angles between vectors are preserved as we translate a space parallel to itself. Yet there is no guarantee that for long closed paths a reflection is not allowed as well as a rigid rotation. A two-dimensional example of such reflections (i.e., a non-orientable surface) occurs in a Möbius strip (Figure 9.3). If we take two vectors, one parallel, one perpendicular to the centerline of a Möbius strip, and go once around starting to the left of the vertical dotted line in Figure 9.3,

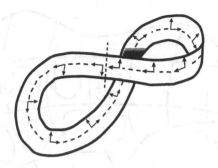

Figure 9.3

the space does not come back upon itself; it has suffered a reflection due to the "twist" in the surface and not simply a rotation.

Now that we have defined the notion of displacement of a vector parallel to itself, we may compute an actual formula for the curvature tensor by going about the path ABCD in Figure 9.2. The differences in the vector at each infinitesimal displacement are given by the Christoffel symbols Γ. But since these are not exactly the same along (AB) and (CD), even though the displacements are the negatives of each other, the vector does not return to its previous value. We can understand how the gradients of the Christoffel symbols get into the act. Carrying out the algebra, we arrive at the expression in eq.(9.2.14).

The curvature tensor itself may be shown to satisfy the Bianchi identity

$$R^{\mu}{}_{\sigma\alpha\beta;\gamma} + R^{\mu}{}_{\sigma\beta\gamma;\alpha} + R^{\mu}{}_{\sigma\gamma\alpha;\beta} = 0. \qquad (9.3.7)$$

I do not, offhand, know the geometrical significance of the Bianchi identity. There is a familiar equation of electrodynamics which can be put in a form which is identical except for the number of dimensions. The field tensor is given in terms of the vector potential by

$$F_{\mu\nu} = \frac{\partial A_{\mu}}{\partial x^{\nu}} - \frac{\partial A_{\nu}}{\partial x^{\mu}}, \qquad (9.3.8)$$

in other words, $F_{\mu\nu}$ is the curl of some vector. But the characteristics of $F_{\mu\nu}$ contained in the statement that $F_{\mu\nu}$ is a curl are equally well described by the identity

$$F_{\mu\nu,\sigma} + F_{\nu\sigma,\mu} + F_{\sigma\mu,\nu} = 0, \qquad (9.3.9)$$

which is of a form like that of the Bianchi identity. The characteristics of a curl may be related to a path integral by means of Stokes' Theorem,

$$\oint_{\Gamma} \vec{G} \cdot d\vec{r} = \int (\text{curl } \vec{G}) \cdot d\vec{S}, \qquad (9.3.10)$$

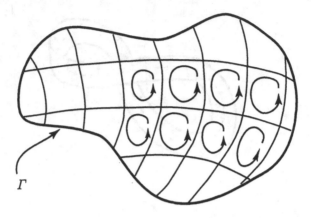

Figure 9.4

where the second integral is a surface integral over any surface bounded by the path Γ.

In the case of gravitation, the analogy might be the following: The path integral represents the change in a vector as we displace it, keeping it parallel to itself, along a closed path Γ. This total change is possibly related to an integral over any two-dimensional hyper-surface bounded by Γ. The proof of such a relation may be looked for in analogy to the proof of Stokes' theorem, which considers a finite surface as divided into an infinitesimal mesh, for example, as illustrated in Figure 9.4; the sum of the contributions of each infinitesimal mesh is shown to be equivalent to the path integral. When viewed as a higher-dimensional analogy of this situation, we may understand better the role of the Bianchi identity in describing the nature of the curvature of space.

9.4   THE CONNECTION BETWEEN CURVATURES AND MATTER

We have seen how the effects of gravity fields may be described in our geometrical interpretation in terms of the curvature tensor $R^{\mu}{}_{\nu\sigma\tau}$. The only remaining task is to connect the curvature tensor to the sources of gravity, matter and energy. The first thing that we do is to contract the first and last indices of the curvature tensor to obtain the Ricci tensor.

$$R_{\nu\sigma} = R^{\mu}{}_{\nu\sigma\mu} \tag{9.4.1}$$

This is the only way to contract the curvature tensor once. The next hint comes from the generalized law of energy and momentum conservation, which says that the contracted covariant derivative, that is, the covariant divergence of the stress-energy tensor, must be zero.

We search for a form involving the Ricci tensor in such a way that its contracted covariant derivative is identically zero. The answer comes from contracting the Bianchi identity, eq.(9.3.7), twice. The contraction in the indices $(\mu\beta)$ results in an expression involving the Ricci tensors.

$$R_{\sigma\alpha;\gamma} - R_{\sigma\gamma;\alpha} + R^{\mu}{}_{\sigma\gamma\alpha;\mu} = 0 \qquad (9.4.2)$$

Contracting now in (σ, α) we get

$$R_{;\gamma} - R^{\sigma}{}_{\gamma;\sigma} - R^{\mu}{}_{\gamma;\mu} = 0. \qquad (9.4.3)$$

So that the tensor quantity which has zero covariant divergence is

$$\left(R^{\mu}{}_{\gamma} - \frac{1}{2} g^{\mu}{}_{\gamma} R \right)_{;\mu} = 0. \qquad (9.4.4)$$

It was Einstein's guess that this quantity is precisely the stress-energy tensor. To write this in Einstein's form, we simply raise one index to write a doubly contravariant tensor.

$$G^{\mu\nu} = R^{\mu\nu} - \frac{1}{2} g^{\mu\nu} R = \lambda^2 T^{\mu\nu}$$
$$G^{\mu\nu}{}_{;\nu} = 0 \qquad (9.4.5)$$

Eq.(9.4.5) is the complete law of gravitation of Einstein; the starting point of all our work. $G^{\mu\nu}$ is often called the Einstein tensor.

An interesting question arises after we have made this connection between the stress-energy tensor and the curvature tensor. Our intuition would assume that if throughout space there is no matter and no stresses, the geometry would be flat, with the Minkowski metric of special relativity. However, it has been possible to find solutions such that

$$T_{\mu\nu} = 0 \quad \text{everywhere, and yet} \quad R^{\sigma}{}_{\rho\tau\nu} \neq 0. \qquad (9.4.6)$$

The most interesting of these is due to A. Taub—it is the most interesting, because it is time-independent; there may be yet other solutions to this problem, so we may well ask, can we have gravity without having sources?

The answer is likely to be similar to the answer we would give to a similar question in electrodynamics. If time dependences are allowed, the equations permit the presence of fields without sources (i.e., traveling waves), yet we have never encountered any physical difficulties in assuming that all observed radiation does come from charged sources which emitted them. It is possible to construct static fields, for example, having potentials

$$\phi = x,$$
$$\phi = x^2 + y^2 - 2z^2, \qquad (9.4.7)$$

which are divergenceless, hence have no sources. But the usual interpretation has been that such fields are due to charges lying outside a certain volume inside which eq.(9.4.7) are valid, and it takes an increasingly large amount of outside charge to make solutions of this kind be reasonable as we attempt to increase the volume over which these solutions hold.

Without having checked in detail the solution of A. Taub, I suspect that he has encountered a similar situation. In order to explain the matterless curvatures, we must take the limiting case of solutions which have a clear physical interpretation over small regions, and allow these regions to become infinitely large—the price that must be paid is to postpone indefinitely the explanation of the growing amount of "outside" matter that is required.

10.1 THE FIELD EQUATIONS OF GRAVITY

We have discovered a tensor, called the curvature tensor, defined by what happened as we carried vectors about in our space. Since it is a tensor, we can use it to generate quantities to be used in writing down covariant equations. We get no physics by just writing down covariant equations, however, we must specify the connection of the equations to the real world of matter. What Einstein did was simply to guess what the connection was. There is no way to deduce the connection from more fundamental principles. Each possible guess has its characteristic properties, hence it is possible for a later worker to assume some criterion which would have made the choice unique, but this is cheating.

The physical hint that may help us is that gravity couples to the energy density, so that, since the energy density in relativity is the (44) component of a second rank tensor, we need a second rank tensor in the equations. The curvature is a fourth rank tensor, so that we contract it once and use the Ricci tensor. It was Einstein's first guess that the stress-energy tensor was simply equal to the Ricci tensor, $\lambda^2 T_{\mu\nu} = R_{\mu\nu}$. However, another choice is possible; we can add to the Ricci tensor a tensor multiple of the scalar curvature R (the contracted Ricci tensor). This is what Einstein finally chose:

$$R_{\mu\nu} - \frac{1}{2}g_{\mu\nu}R = \lambda^2 T_{\mu\nu}. \tag{10.1.1}$$

There is good reason why this choice is better. If we take the covariant divergence of eq.(10.1.1), the answer is identically zero. This means that the law of conservation of energy is a consequence simply of the form of the equation (10.1.1). If we had set the stress-energy tensor equal to the Ricci tensor alone, the law of energy conservation would have been a physical postulate, an additional requirement which would bring in more information and which would result in less freedom. As it is, the metric tensors are not unique; in dealing with them, we have the freedom to choose four functions, corresponding to four functions which describe a general transformation giving the new coordinates in terms of the old ones. Since the law of energy conservation is an identity, four functions in the metric are truly free.

How well does Einstein's choice fit Nature? How do we get $T^{\mu\nu}$, and what is the significance of the equations and of the curvatures? In order to answer these questions, we shall play with the equations for a while. First of all, we will try to understand the relation to the rest of physics and to variational principles.

To write down an action principle relativistically, we need an integral which is a scalar invariant. We choose the action for the gravitational field to be

$$S_g = -\frac{1}{2\lambda^2} \int d^4x\, R\sqrt{-g}. \tag{10.1.2}$$

In what follows, we abbreviate $d^4x = dx\,dy\,dz\,dt$. The action S_g is a scalar because R is a scalar and $\sqrt{-g}\,d^4x$ is a scalar. We can show the latter from considering that the proper time is a scalar invariant.

$$(ds)^2 = g_{\mu\nu}\,dx^\mu\,dx^\nu \tag{10.1.3}$$

In view of the fact that $g_{\mu\nu}$ is a symmetric tensor, we can always introduce a rotation so that it is diagonal; in this case,

$$(ds)^2 = D(dt)^2 - C(dz)^2 - B(dy)^2 - A(dx)^2. \tag{10.1.4}$$

From this we see that the volume element d^4x is not an invariant; the properly invariant volume element is

$$\sqrt{ABCD}\,dx'\,dy'\,dz'\,dt' = \sqrt{-g'}\,d^4x', \tag{10.1.5}$$

where $g' = \text{Det} \, g'_{\mu\nu}$. Under orthogonal transformations, then $d^4x = d^4x'$, and also $\text{Det} \, g_{\mu\nu}$ equals $\text{Det} \, g'_{\mu\nu}$. Thus, the general expression for the invariant volume element is

$$\sqrt{-g} \, d^4x. \tag{10.1.6}$$

The quantity $\sqrt{-g}$ is a scalar density. This means that its change under a coordinate transformation is obtained by multiplication by the Jacobian of the transformation.

$$\sqrt{-g'} = \left| \frac{\partial x^\mu}{\partial x'^\nu} \right| \sqrt{-g} \tag{10.1.7}$$

The curvature tensor appears when we take the variation of S_g with respect to $g_{\mu\nu}$.

$$\frac{\delta S_g}{\delta g_{\mu\nu}} = \frac{1}{2\lambda^2} \sqrt{-g} \left(R^{\mu\nu} - \frac{1}{2} g^{\mu\nu} R \right) \tag{10.1.8}$$

It is because of this that we can use the integral of R as the action of the gravitational part of the complete problem.

Let me point out that because this stress tensor appears in this way, from a variational principle, its covariant divergence is necessarily zero (a point first emphasized, I believe, by Eddington). We have seen the connection from the other direction—that we could deduce a variational principle provided that we started from a divergenceless tensor. The proofs that we offer are not rigorous; we don't bother with the rigor because it is the facts that matter, and not the proofs. Physics can progress without the proofs, but we can't go on without the facts. The proofs are useful in that they are good exercises; if the facts are right, then the proofs are a matter of playing around with the algebra correctly.

We want to show that if the functional

$$S_g = \int d^4x \, \Sigma[g_{\mu\nu}] \tag{10.1.8'}$$

is invariant under coordinate transformations, then the covariant divergence of the variation of S_g with respect to $g_{\mu\nu}$ is identically zero. Under the infinitesimal transformation to primed coordinates, $x^\mu \to x'^\mu$,

$$x^\mu = x'^\mu + h^\mu(x'), \tag{10.1.9}$$

the change of $g_{\mu\nu}$ is given by

$$g_{\mu\nu}(x) \to g'_{\mu\nu}(x') = g_{\mu\nu}(x') + h^\alpha{}_{,\nu} g_{\mu\alpha}(x') + h^\alpha{}_{,\mu} g_{\nu\alpha}(x') + h^\alpha g_{\mu\nu,\alpha}(x'). \tag{10.1.10}$$

In terms of the new coordinates, dropping the primes on the integration variables, the invariant action is

$$S_g = \int d^4x \, \Sigma[g'_{\mu\nu}]$$

$$= \int d^4x \, \Sigma[g_{\mu\nu}] + \int d^4x \, \frac{\delta\Sigma}{\delta g_{\mu\nu}} \left(h^\alpha{}_{,\nu}g_{\mu\alpha} + h^\alpha{}_{,\mu}g_{\nu\alpha} + h^\alpha g_{\mu\nu,\alpha}\right).$$

$$(10.1.11)$$

When we do an integration by parts on the second term of this expression, we convert it to an expression involving the functional derivatives of the function Σ. We set it equal to zero, since we know that the change in the action must be zero for any h^α, because of the character of Σ.

$$\frac{\partial}{\partial x^\mu}\left[\frac{\delta\Sigma}{\delta g_{\mu\nu}} g_{\nu\alpha}\right] - \frac{1}{2}\frac{\delta\Sigma}{\delta g_{\mu\nu}}\frac{\partial g_{\mu\nu}}{\partial x^\alpha} = 0 \qquad (10.1.12)$$

Let us denote by $\mathcal{G}^{\mu\nu}$ the variation of $2\lambda^2 S_g$ with respect to $g_{\mu\nu}$:

$$\mathcal{G}^{\mu\nu} = 2\lambda^2 \frac{\delta S_g}{\delta g_{\mu\nu}}. \qquad (10.1.13)$$

The quantity $\mathcal{G}^{\mu\nu}$ is a contravariant tensor *density* of second rank. With this definition, equation (10.1.12) becomes

$$(g_{\alpha\mu}\mathcal{G}^{\mu\nu})_{,\nu} - \frac{1}{2}g_{\mu\nu,\alpha}\mathcal{G}^{\mu\nu} = 0, \qquad (10.1.14)$$

which is equivalent to the statement that the covariant divergence of $\mathcal{G}^{\mu\nu}$ is zero.

$$\mathcal{G}^{\mu\nu}{}_{;\nu} = 0 \qquad (10.1.15)$$

In demonstrations of equivalence involving covariant divergences, it is convenient to use some formulae which we shall collect now for later reference. First of all, we compute the value of the contracted Christoffel symbols (Γ's). From the definition,

$$\Gamma^\mu_{\epsilon\mu} = g^{\mu\sigma}[\mu\epsilon,\sigma] = \frac{1}{2}g^{\mu\sigma}[g_{\sigma\mu,\epsilon} + g_{\sigma\epsilon,\mu} - g_{\mu\epsilon,\sigma}]. \qquad (10.1.16)$$

The second and third terms cancel each other because $g_{\mu\nu}$ is symmetric. The remaining term consists of the reciprocal matrix $g^{\mu\nu}$ multiplying the gradient of $g_{\mu\nu}$. From a well-known theorem on determinants, the conjugate minor $M^{\mu\nu}$ of the matrix $g_{\mu\nu}$ is related to the inverse $g^{\mu\nu}$ by

$$g^{\mu\nu} = M^{\mu\nu}/g, \qquad (10.1.17)$$

and thus

$$g_{,\lambda} = g_{\mu\nu,\lambda} M^{\mu\nu} = g_{\mu\nu,\lambda} g^{\mu\nu} g. \tag{10.1.18}$$

Therefore,

$$g^{\mu\nu} g_{\mu\nu,\lambda} = [\log(-g)]_{,\lambda}, \tag{10.1.19}$$

and the contracted Christoffel symbol is

$$\Gamma^{\mu}_{\epsilon\mu} = [\log \sqrt{-g}]_{,\epsilon} = \frac{1}{\sqrt{-g}} (\sqrt{-g})_{,\epsilon}. \tag{10.1.20}$$

Other useful formulae for the covariant divergences are as follows. For a scalar function, covariant gradients are the same as ordinary gradients.

$$\phi_{;\nu} = \phi_{,\nu} \tag{10.1.21}$$

For a contravariant vector, the covariant divergence is

$$A^{\nu}_{;\nu} = \frac{1}{\sqrt{-g}} (\sqrt{-g} A^{\nu})_{,\nu}. \tag{10.1.22}$$

The covariant curl is the same as the ordinary curl.

$$A_{\mu;\nu} - A_{\nu;\mu} = A_{\mu,\nu} - A_{\nu,\mu} \tag{10.1.23}$$

For second rank tensors, the answers are different, depending on the symmetry; for antisymmetric tensors,

$$F^{\mu\nu}_{;\nu} = \frac{1}{\sqrt{-g}} (\sqrt{-g} F^{\mu\nu})_{,\nu}, \quad \text{if} \quad F^{\mu\nu} = -F^{\nu\mu}. \tag{10.1.24a}$$

And, for symmetric tensors

$$T^{\nu}_{\mu;\nu} = \frac{1}{\sqrt{-g}} (T^{\nu}_{\mu} \sqrt{-g})_{,\nu} - \frac{1}{2} g_{\alpha\beta,\mu} T^{\alpha\beta}, \quad \text{if} \quad T^{\mu\nu} = T^{\nu\mu}. \tag{10.1.24b}$$

Using these formulae, it is quite straightforward to deduce that eqs. (10.1.14) and (10.1.15) are equivalent. Thus, we see that the invariance of an action results in the construction of a tensor density which is automatically divergenceless. Since the covariant derivative of the metric tensor $g_{\mu\nu}$ vanishes, the covariant derivative $(\sqrt{-g})_{;\lambda}$ also vanishes. (Note carefully that the ordinary derivative $(\sqrt{-g})_{,\lambda}$ is not the same thing, since $\sqrt{-g}$ is a scalar density and not a scalar.) The tensor associated with the tensor density $\mathcal{G}^{\mu\nu}$ is also divergenceless.

$$G^{\mu\nu} = \mathcal{G}^{\mu\nu}/\sqrt{-g}, \qquad G^{\mu\nu}_{;\nu} = 0. \tag{10.1.25}$$

In this connection, we should like to clarify a point which is small but sometimes tricky. In Lecture 6, we worked with functional equations (e.g., eq.(6.2.3)) of the same form as eq.(10.1.2); the solutions to these equations are really tensor *densities* and not tensors. The stress-energy tensor density $\mathcal{T}^{\mu\nu}$ satisfies the equation

$$\mathcal{T}^{\mu\nu}{}_{,\nu} = -\Gamma^{\mu}_{\alpha\beta}\mathcal{T}^{\alpha\beta}, \tag{10.1.26}$$

where

$$\mathcal{T}^{\mu\nu} = \sqrt{-g}T^{\mu\nu},$$

but the stress-energy tensor $T^{\mu\nu}$ satisfies the following:

$$T^{\mu\nu}{}_{,\nu} = -\Gamma^{\mu}_{\alpha\beta}T^{\alpha\beta} - \frac{1}{2g}g_{,\alpha}T^{\mu\alpha}. \tag{10.1.27}$$

10.2   THE ACTION FOR CLASSICAL PARTICLES IN A GRAVITATIONAL FIELD

Next, we discuss how one writes down a general law of physics, one which describes not only the gravity fields, but also the matter. We assume that it can be deduced from a principle of least action; the mathematical statement is that the variation of the action is zero.

$$\delta S = \delta \int d^4x \, \mathcal{L}[g_{\mu\nu}, A_\mu, \ldots] = 0 \tag{10.2.1}$$

The Lagrangian density $\mathcal{L}$ contains various kinds of fields, for example, the gravity tensor field $g_{\mu\nu}$, the electromagnetic field A_μ, and, if matter is scalar, a scalar matter field ϕ. When we vary this action with respect to the various fields, we get the equations of propagation for the corresponding fields. We have written down one piece of this action; let us denote what is left over by a quantity S_m which depends on the matter fields ϕ and electromagnetic fields A_μ and all other fields that we know of. When we take the variation of

$$S = S_g + S_m = -\frac{1}{2\lambda^2}\int d^4x \, \sqrt{-g} \, R + S_m, \tag{10.2.2}$$

with respect to $g_{\mu\nu}$, we get the following equation:

$$\frac{\delta S_g}{\delta g_{\mu\nu}} = \frac{1}{2\lambda^2}\sqrt{-g}\left[R^{\mu\nu} - \frac{1}{2}g^{\mu\nu}R\right] = -\frac{\delta S_m}{\delta g_{\mu\nu}}. \tag{10.2.3}$$

The stress-energy tensor density of matter $T^{\mu\nu}$ must be the variational derivative of S_m,

$$T^{\mu\nu} = -2\frac{\delta S_m}{\delta g_{\mu\nu}},\tag{10.2.4}$$

if $T^{\mu\nu}$ is to be the source of the gravitational field. We now need some examples for $T^{\mu\nu}$. If we are unable to calculate $T^{\mu\nu}$ by some other physical principle, there is no theory of gravitation, since we do not know how the fields are related to any other object.

There are some consistency requirements similar to those we find in electromagnetism. In order to solve Maxwell's equations, we need to have the currents. They must be conserved currents, not just arbitrary currents. The conserved source currents which are meaningful are obtained by solving some other problems of physics, following some independent law, such as Ohm's Law or Hooke's Law or Schrödinger's equation for such-and-such a system. If we did not have these other laws, the theory of electromagnetic fields would be useless and empty of meaning.

For gravity the thing is more complicated. The tensor $T^{\mu\nu}$ involves the motion of matter, hence we must have a law which matter follows, including Ohm's Law and Hooke's Law; but also $T^{\mu\nu}$ will involve the gravity fields $g_{\mu\nu}$, a circumstance which tangles up the problems much more than in electromagnetism. In general, it is not possible to write down any kind of consistent $T^{\mu\nu}$ except for the vacuum, unless one has already solved the complete, tangled problem. The trouble is that any specified $T^{\mu\nu}$ will not solve the problem except for special cases of the metric tensor $g_{\mu\nu}$; the complete relativistic solution should work regardless of the particular choice of coordinates and their curvatures. Even for very simple problems, we have no idea of how to go about writing down a proper $T^{\mu\nu}$. We do not know how to write a $T^{\mu\nu}$ to represent a rotating rod, so that we cannot calculate exactly its radiation of gravity waves. We cannot calculate the $T^{\mu\nu}$ for a system consisting of the earth and the moon, because the tidal forces and the elasticity of the earth change the gravity fields significantly. If we assume that the earth is rigid, the equations are inconsistent. If we assume that the earth is a point, the equations are too singular to have solutions. And yet it is obvious that a glob of matter of a given stiffness, such as the earth, will rotate about a moon of another mass and stiffness, whether or not the equations are manageable.

The theory of gravity suffers at this point because one side of the equation is beautiful and geometric, and the other side is not—it has all the dirt of Hooke's Law and of the other laws that govern matter, and these are neither pretty nor geometric. Many physicists have become so hypnotized by the beauty of one side of the equation that they ignore the other, and hence have no physics to investigate.

We have to learn to guess at forms for the action term S_m. As a starting point, it is useful to consider classical limits. If we write down correct classical actions, it is usually not very hard to see how to generalize the formula so that it becomes invariant under arbitrary coordinate transformations. A convenient method to generate such generalized formulae is to go back to the locally falling (freely falling) tangent coordinate system, and figure out how to add in factors of $g_{\mu\nu}$ and $R^{\mu\nu}$ to make the thing invariant. For example, a free particle under no forces has an action

$$S_m = -\frac{m_0}{2} \int ds \, \frac{dz^\mu}{ds} \frac{dz_\mu}{ds}. \tag{10.2.5}$$

This example illustrates the procedure for guessing; it has usually been found to be fruitful. We write down things as they are in flat coordinates, change to curvilinear coordinates, and see where the $g_{\mu\nu}$'s come in. It is often obvious which general form will reduce to the flat space results. If $z^\mu(s)$ is the orbit of a particle, freely falling, the term S_m of the action is

$$S_m = -\frac{m_0}{2} \int d^4x \, ds \, \delta^4(x - z(s)) \, g_{\mu\nu} \frac{dz^\mu}{ds} \frac{dz^\nu}{ds}. \tag{10.2.6}$$

The stress-energy tensor density $T^{\mu\nu}$ is obtained by variation of this S_m with respect to $g_{\mu\nu}$, which gives

$$T^{\mu\nu} = m_0 \int ds \, \delta^4(x - z(s)) \frac{dz^\mu}{ds} \frac{dz^\nu}{ds}. \tag{10.2.7}$$

The analogy to results in electromagnetism is so strong, that this result is not surprising. There are no troubles of inconsistency with this $T^{\mu\nu}$; since we have started from an invariant form, this $T^{\mu\nu}$ satisfies the covariant divergence condition.

It is interesting to examine the relation between the equations of motion and the divergenceless stress-energy tensor from an inverse point of view. In writing the action S_m we have essentially asserted that the particle moves along a geodesic. The resulting stress-energy tensor density is thereby divergenceless. We want now to show the reverse. Suppose that $T^{\mu\nu}$ is nonzero only in a filamentary region of space-time. Then, we can show that the filamentary region is indeed a geodesic, provided only that we assume something equivalent to a spherical symmetry of the particle as we see it at very close range. The hint is to start from the condition that relates the ordinary divergence of $T^{\mu\nu}$ to $T^{\mu\nu}$ itself, and perform an integration by parts to convert the volume integration to a surface integration.

$$\int d^4x \, T_\mu{}^\nu{}_{,\nu} = \frac{1}{2} \int d^4x \, g_{\alpha\beta,\mu} \, T^{\alpha\beta} \tag{10.2.8}$$

If $T^{\mu\nu}$ has no value except over a filamentary region, the contributions to the surface integral vanish except where the filament intersects the surface—and these correspond to the momenta of the particle "before" and "after" if these surfaces are at constant time. Converting this result to differential form eventually results in showing that the motion follows the geodesic equation:

$$\frac{d^2z^\mu}{ds^2} + \Gamma^\mu_{\alpha\beta}\frac{dz^\alpha}{ds}\frac{dz^\beta}{ds} = 0. \tag{10.2.9}$$

The possibility of this deduction leads to the statement that the Einstein equations simultaneously determine the motion of matter and the gravity fields. This statement is misleading and is not quite as remarkable as it may seem at first. Let us recall that if we have a free particle all by itself, far away from anything else, then the laws of energy and momentum conservation determine its motion completely. In gravity theory, a freely falling particle becomes equivalent to a free particle, so that again energy conservation is enough to determine the motion completely. But the usual physical situation is not as simple as this. For when we have more than just gravity and a particle, the equations of motion do not follow from the laws of conservation of energy and momentum alone. In electrodynamics, the conservation of charge must hold in any solution of Maxwell's equations, so it may be said to be a consequence of the equations. But this does not serve all by itself to construct the equations of motion for the charges, the fields they produce, and the forces they exert upon each other. Likewise, in gravitation theory the conservation of energy and momentum hold, but this does not suffice to determine the motion of the planets and the moon, for they are not points, and laws of physics other than the conservation of energy are required to elucidate their behavior in a gravitational field.

10.3   THE ACTION FOR MATTER FIELDS IN A GRAVITATIONAL FIELD

Next, we prepare the passage to quantum theory. If scalar particles are described by a scalar field ϕ, then the appropriate action term in flat coordinates is

$$S_m = \frac{1}{2}\int d^4x\,(\phi_{,\nu}\phi^{,\nu} - m^2\phi^2). \tag{10.3.1}$$

The generalization to curvilinear coordinates may be easily made; we guess that

$$S_m = \frac{1}{2}\int d^4x\,\sqrt{-g}\,(g^{\mu\nu}\phi_{,\mu}\phi_{,\nu} - m^2\phi^2). \tag{10.3.2}$$

This form is evidently invariant under arbitrary coordinate transformations, which is one requirement, and it reduces to the flat result in flat space. We can, however, write down other terms which are perfectly good invariants, quadratic in the fields ϕ, involving the curvature tensor. These terms all go to zero as the space becomes flat. It is conceivable that the action should contain proportions α and β of the following terms, for example

$$-\alpha \int d^4x \sqrt{-g}\, R\phi^2 + \beta \int d^4x \sqrt{-g}\, (R^{\mu\nu}\phi_{,\mu}\phi_{,\nu})\,. \qquad (10.3.3)$$

We see that the action we write down is not unique. The first term we have written down must be there, since it is the only one that reduces to the flat result. And there is no experimental evidence on the tidal forces or what-not which might be a criterion for the inclusion or noninclusion of other terms such as in eq.(10.3.3). The only reasonable thing a physicist can do now is to choose some of these terms as being "simpler" than others, disregard the complicated ones, and see what kind of theory he has left. In some kind of sense, perhaps derivatives are more complicated things than just fields, so that the β term is more complicated because it contains four derivatives, two in the fields and two in the $R^{\mu\nu}$. The α term has only two, however, both on the $g_{\mu\nu}$ field. But it is difficult to define complication in an unambiguous way; it is always possible to perform integrations by parts so that the derivatives disappear in one place and reappear in another—the simplicity which is apparent in starting from one point may not correspond to the simplicity which would result from a different starting point. If we are used to starting quantum mechanics from Schrödinger's equation, the simplest action seems to be that for which $\alpha = 0$. But had we started from quantum mechanics given in terms of path integrals, the simplest action would seem to be that for which $\alpha = 1/6$. Each of these appear to be the simplest choice of α from their respective viewpoints. I do not know of any satisfactory way to determine α, and feel that the action for a scalar field is ambiguous.[*]

The significance of a term like the β one in eq.(10.3.3) has to do with whether a particle can feel the gravity field over a region large enough to sense the local curvature. If the particle has a structure which is somehow infinitesimally small, then it cannot feel the curvature. But if rather it performs a corkscrew motion about its position, as it moves about, a term involving the local curvature might very well be present.

We give an example of how a different starting point yields a different answer in an innocent way, by considering a situation in electrodynamics. There the principle of minimal electromagnetic coupling leads to the

---

[*] For a modern perspective involving the hydrogen spectrum, see [Klei 89].

replacement

$$\frac{\partial}{\partial x^\mu} \longrightarrow \left(\frac{\partial}{\partial x^\mu} - ieA_\mu \right) \qquad (10.3.4)$$

in the Lagrangian. Suppose now that, before we make this replacement, we had written a Lagrangian integral as follows:

$$S = \int dV \, \overline{\psi} \gamma^\mu \frac{\partial}{\partial x^\mu} \psi - \int dV \, \overline{\psi} m \psi + \epsilon \int dV \, \overline{\psi} (\gamma^\mu \gamma^\nu - \gamma^\nu \gamma^\mu) \frac{\partial}{\partial x^\mu} \frac{\partial}{\partial x^\nu} \psi.$$
$$(10.3.5)$$

The last term is not written down in the usual formulation, because it is identically zero; but precisely because it is identically zero, there can't be any hard and fast rule about leaving it in! However, when we make the replacement of the gradient by eq.(10.3.4) to include electromagnetism, the resulting Lagrangian is not the same; it has a new term,

$$\epsilon \gamma^\mu \gamma^\nu F_{\mu\nu}, \qquad (10.3.6)$$

where

$$F_{\mu\nu} = A_{\nu,\mu} - A_{\mu,\nu}.$$

This term is the anomalous moment term, discovered by Pauli. (This was first pointed out to me by Wentzel.)

The electrodynamics of particles of spin 1 is complicated also by anomalous quadrupole moments. There is no obviously simpler Lagrangian to write, so that theoretical papers must present calculations with alternative theories which correspond to different anomalous moments.

In our gravity theory, the situation is analogous. It is as though the particle had an anomalous moment of inertia, added to the usual one due to a mass distribution.

In electromagnetism, such ambiguities do not appear in the description of spin-zero particles—they first appear in discussing spin-1/2 particles. On the other hand, in gravity, the difficulties appear in discussing even the simplest case of scalar particles. There is no solution to these difficulties—we have to recognize that many alternative theories (different values of α) are possible.

The motion of a particle in a given gravity field is described by the equation which results when we vary the action with respect to the field ϕ. Depending on how we convert to quantum mechanics, different actions lead to the simplest results. Thus, we can't argue that something is simpler unless it leads to a simultaneous simplicity in the solution of many different problems. For different problems, different values of α are needed in simplifying, or in getting rid of divergences. Setting $\alpha = 0$ therefore results in a covariant simplicity only in the sense that less algebra is required in some calculations at the start. There is no implicit physical simplicity, since all values of α lead to various degrees of trouble in one problem or another.

Let us proceed to obtain the equations of motion of the matter field ϕ. Starting from eq.(10.3.2), we may use the following variations of the reciprocal matrix and square root of the determinant,

$$\delta g^{\mu\nu} = -g^{\mu\alpha}g^{\nu\beta}\,\delta g_{\alpha\beta}, \qquad \delta(\sqrt{-g}) = \frac{1}{2}\sqrt{-g}g^{\alpha\beta}\,\delta g_{\alpha\beta}, \qquad (10.3.7)$$

to get the following expression for $T^{\mu\nu}$:

$$T^{\mu\nu} = -2\frac{\delta S_m}{\delta g_{\mu\nu}} = \sqrt{-g}\phi^{;\mu}\phi^{;\nu} - \frac{1}{2}\sqrt{-g}g^{\mu\nu}(\phi_{;\alpha}\phi^{;\alpha} - m^2\phi^2)$$

$$- \alpha\sqrt{-g}\left(R^{\mu\nu} - \frac{1}{2}g^{\mu\nu}R\right)\phi^2 - 4\alpha\phi\phi_{,\beta}\sqrt{-g}\frac{\delta R}{\delta g_{\mu\nu,\beta}} \qquad (10.3.8)$$

Next, we vary with respect to the field ϕ and set the variation equal to zero, to obtain something which is an analogue of the Klein-Gordon equation.

$$\left(\sqrt{-g}g^{\mu\nu}\phi_{,\nu}\right)_{,\mu} + \sqrt{-g}m^2\phi + 2\alpha R\sqrt{-g}\phi = 0 \qquad (10.3.9)$$

We obtain an equation in which tensors appear by dividing through by the scalar density $\sqrt{-g}$.

$$\frac{1}{\sqrt{-g}}\left(\sqrt{-g}g^{\mu\nu}\phi_{,\nu}\right)_{,\mu} + m^2\phi + 2\alpha R\phi = 0 \qquad (10.3.10)$$

Using eqs.(10.1.20) and (10.1.21) we see that this becomes

$$\phi_{;\mu}{}^{\mu} + (m^2 + 2\alpha R)\phi = 0. \qquad (10.3.11)$$

The relation to the Klein-Gordon equation may be seen by considering the case $\alpha = 0$; the ordinary D'Alembertian has simply been replaced by its covariant analogue, the covariant D'Alembertian.

The preceding steps have given us a definite theory, since we have specified precisely how matter moves and what the source tensor is. It is easy to check that for the tensor we have written, the covariant divergence $T^{\mu\nu}{}_{;\nu}$ is zero—the trick is to use whenever necessary, the field equations themselves. Thus, all is consistent, all proper tensors are divergenceless, and it is meaningful to proceed from here. In order to construct a more complete theory, we add terms to the action so as to represent all the other known fields. We write the action first in flat space, as we know it, choose the simplest form from some kind of criterion which is invariant. The requirement that the action should be invariant results in covariant equations for the fields. This is not a restriction on what known fields we can include, because all the known laws of physics can be covariantly written. Differential laws have this property. Any law written as a differential equation may easily be converted to a covariant form; we assume that in the tangent space the law is the same as the one we know, and then rotate and stretch the coordinates. The resulting equations involve derivatives of the fields and up to two derivatives of the metric tensor.

Once we have written all processes, first in differential form, then in covariant form, then we can use our theory to compute, for example, the motion of matter in a star. The processes considered may be described by opacity laws, the laws of scattering, etc. What is not allowed is to use laws which would violate energy conservation. We may not, for example, say good bye to neutrinos which are created; the neutrinos lose energy, because of the gravitational potential, as they leave the star, and a consistent theory cannot be written if we neglect this, and the effect of their energy density in modifying the gravitational field. Therefore, it will not suffice to write integral diffusion equations with finite mean free paths, but we must follow the motions of the diffusion particles by the complete laws in the form of differential equations.

To make things easier for us later, we write down here the integrand of the action for the fields directly in terms of the metric tensor. Our previous expressions look simpler because they are defined in terms of combinations of the metric tensor, but this form is often much more useful.

$$-R\sqrt{-g} =$$
$$-\frac{\sqrt{-g}}{4}\, g_{\nu\lambda,\sigma} g_{\mu\rho,\tau} \left(g^{\nu\lambda} g^{\sigma\tau} g^{\mu\rho} - g^{\nu\mu} g^{\lambda\rho} g^{\sigma\tau} + 2g^{\nu\mu} g^{\lambda\tau} g^{\sigma\rho} - 2g^{\tau\mu} g^{\rho\sigma} g^{\nu\lambda}\right)$$
$$+ \left[\sqrt{-g}\, g_{\nu\sigma,\mu}(g^{\sigma\nu} g^{\rho\mu} - g^{\mu\nu} g^{\rho\sigma})\right]_{,\rho} \qquad (10.3.12)$$

The last term is a divergence, and it integrates to zero in the action, so often we may leave it out of the considerations altogether. For many problems, it will be sufficient to write down the action as the integral of the first term, denoted by H, so that

$$\delta S_g = -\frac{1}{2\lambda^2}\, \delta \int d^4x\, H,$$

where

$$H = \sqrt{-g}\, g^{\mu\nu} \left[\Gamma^\rho_{\nu\sigma}\Gamma^\sigma_{\rho\mu} - \Gamma^\rho_{\mu\nu}\Gamma^\sigma_{\rho\sigma}\right]. \qquad (10.3.13)$$

We are now ready again to make a quantum theory, after having a theory from Einstein's point of view. The theory is more complete than when we were discussing the Venutian viewpoint—we have the complete Lagrangians including interaction with matter correct to all orders. If we restrict our attention to a universe consisting only of gravity fields and

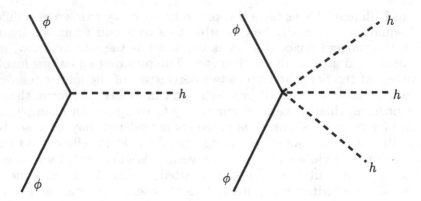

Figure 10.1

scalar matter, the field theory is obtained by considering expansions in terms of a coupling constant:

$$g_{\mu\nu} = \eta_{\mu\nu} + 2\lambda h_{\mu\nu}. \tag{10.3.14}$$

In the Lagrangian, the terms which are quadratic correspond simply to the propagators, the terms involving products of two ϕ and one h, and terms involving three h's and two ϕ's, correspond to diagrams such as are shown in Figure 10.1. In this way, we have arrived at a prescription for calculating quantum mechanical amplitudes for the motion of matter after having started out from a geometrical point of view.

In the time to come, we shall use the classical theory to discuss the motion of classical objects and to discuss cosmological questions, and we shall use the quantum theory to compute the radiation of gravitational waves. The third alternative viewpoint on gravity will be presented after we have discovered ways in which the quantum mechanical theory gets us into hot water.

In considering the terms in the action, we might consider why the field term might not include a certain proportion Λ of $\int d^4x \sqrt{-g}$. This would be an integral proportional to the volume of the universe, presumably a constant. The resulting equation for the field behaves somewhat as though the gravitons had a mass and a universal source. The observation of the extremely long range of gravity forces makes it rather pointless to introduce such a term, even though it might lead to a consistent theory. The equations of motion that come from it are

$$G_{\mu\nu} = \Lambda g_{\mu\nu} + \lambda^2 T_{\mu\nu}. \tag{10.3.15}$$

The constant Λ is known as the "cosmological constant." Einstein wanted a closed universe, so he assigned to it a value which made steady-state solutions of such a universe possible. This he later referred to as his Great Mistake; had he chosen it equal to zero, he would have concluded that the universe must be expanding (or contracting). It was only later that Hubble discovered that the faraway galaxies were receding and the universe was expanding. Ever since this change in Einstein's theory of the universe was introduced, cosmology has been hounded by difficulties in the determination of the cosmological constant. I agree with Einstein's second guessing and think $\Lambda = 0$ is most likely.

11.1 THE CURVATURE IN THE VICINITY OF A SPHERICAL STAR

We shall now turn our attention to finding actual solutions to the Einstein equations for some cases of physical interest. As it turns out that there are very few observations about gravity which cannot be adequately explained by Newtonian gravitation, there are only two solutions of Einstein's equations which have been sought. One of these is that of the gravity field in the vicinity of a star (which should give correctly the deflection of light and the precession of the orbit of Mercury). The other deals with nearly uniform mass distribution, and it is the one of interest in considering cosmological models.

If we assume spherical symmetry, we expect that the metric tensor would result in an expression possibly of the following form, for the square of an interval of proper time

$$(ds)^2 = A(dt)^2 + B dr\, dt - C(dr)^2 - D\Big((d\theta)^2 + \sin^2\theta\,(d\phi)^2\Big)r^2 \quad (11.1.1)$$

where the symbols A, B, C, D stand for functions which may depend on the coordinates (r, t) but not on (θ, ϕ). This expression allows dynamic solutions in which the motion of matter is purely radial.

It is possible to reduce the number of unknown functions by a judicious choice of new coordinates. For example, if we change the scale of the coordinate r according to the rule

$$r' = \sqrt{D(r,t)}\, r, \qquad\qquad (11.1.2)$$

the resulting expression for $(ds)^2$ in terms of r' and dr' instead of r and dr is of the same form, but the new function D is exactly $D = 1$. So that the function D is superfluous, since $D = 1$ corresponds to a problem of equivalent generality.

A second reduction is achieved by changing the scale of the time. We let

$$t' = t'(t,r). \qquad\qquad (11.1.3)$$

This introduces a new function which may be so chosen that the product $dr\,dt'$ has coefficient zero. This means that by setting $B = 0$ we lose no generality.

It is customary from this point onward, to work, not with the functions A and C, but with new functions ν and λ which are defined as follows:

$$A = e^{\nu}, \qquad C = e^{\lambda}, \qquad\qquad (11.1.4)$$

(this convention follows Schwarzschild). The metric tensor is diagonal, and with the identification of the indices (1,2,3,4) with the coordinates (r, θ, ϕ, t), the components of the metric tensor are the following:

$$g_{44} = e^{\nu}, \qquad g_{11} = -e^{\lambda}, \qquad g_{22} = -r^2, \qquad g_{33} = -r^2 \sin^2 \theta. \quad (11.1.5)$$

Because the tensor is diagonal, the elements of the reciprocal tensor $g^{\mu\nu}$ are the reciprocals of the elements of $g_{\mu\nu}$; explicitly,

$$g^{44} = e^{-\nu}, \qquad g^{11} = -e^{-\lambda}, \qquad g^{22} = -1/r^2, \qquad g^{33} = -\frac{1}{r^2 \sin^2 \theta}.$$
$$(11.1.6)$$

The computation of the elements of the curvature tensor may now be carried out. It is very straightforward although tedious, because there are so many derivatives in the Christoffel symbols and so many sums to be made.

When all is done, the components of the curvature tensor may be expressed in terms of the functions ν and λ and their derivatives with respect to the time t and the radial distance r. In order to write more economically, we use primes and dots to denote the derivatives, as follows:

$$\nu' = \frac{\partial \nu}{\partial r}, \qquad \dot{\nu} = \frac{\partial \nu}{\partial t}, \qquad \text{etc.} \qquad (11.1.7)$$

The explicit expressions are the following.

$$R^{42}{}_{41} = -e^{-\lambda}\left(\frac{1}{2}\nu'' + \frac{1}{4}(\nu')^2 - \frac{1}{4}\lambda'\nu'\right) + e^{-\nu}\left(\frac{1}{2}\ddot{\lambda} + \frac{1}{4}(\dot{\lambda})^2 - \frac{1}{4}\dot{\lambda}\dot{\nu}\right)$$

$$R^{42}{}_{42} = R^{43}{}_{43} = -\frac{1}{2r}\nu'\,e^{-\lambda}$$

$$R^{21}{}_{21} = R^{31}{}_{31} = \frac{1}{2r}\lambda'\,e^{-\lambda}$$

$$R^{32}{}_{32} = -\frac{1}{r^2}\left(e^{-\lambda} - 1\right)$$

$$R^{42}{}_{12} = R^{43}{}_{13} = -\frac{1}{2r}\dot{\lambda}e^{-\nu}$$

$$(11.1.8)$$

All other components are zero, unless they may be obtained by a trivial permutation of indices of some element in eq.(11.1.8).

11.2   ON THE CONNECTION BETWEEN MATTER AND THE CURVATURES

It is the tensors reduced from the curvature tensor which are related to the stress-energy tensor. The combinations involved are

$$G^{\mu}{}_{\nu} = R^{\mu}{}_{\nu} - \frac{1}{2}g^{\mu}{}_{\nu}R. \qquad (11.2.1)$$

The components of the tensor $G^{\mu}{}_{\nu}$ have rather simple expressions in terms of sums of the elements of $R^{\mu\nu}{}_{\sigma\tau}$. For example, the diagonal elements are

$$G^4{}_4 = R^{12}{}_{12} + R^{13}{}_{13} + R^{23}{}_{23},$$
$$G^1{}_1 = R^{42}{}_{42} + R^{43}{}_{43} + R^{32}{}_{32}. \qquad (11.2.2)$$

In other words, each involves a sum over those elements of $R^{\mu\nu}{}_{\sigma\tau}$ which do not involve the diagonal index. For the off-diagonal elements we also get very simple expressions. For example,

$$G^4{}_1 = R^{24}{}_{12} + R^{34}{}_{13},$$
$$G^2{}_1 = R^{32}{}_{13} + R^{42}{}_{14}, \qquad (11.2.3)$$

and by analogy with these, we can write quite simply the appropriate expressions for other components.

The simplicity of the sums involved can suggest to us an interpretation of the curvatures in terms of matter distributions. We have previously discussed the curvature of a two-dimensional surface in terms of the fractional defect of the circumference or the area of a circle, with respect to its value in flat-space in terms of the measured radii:

$$\text{circumference} = 2\,\pi r(1 - \text{coeff} \times \text{area}). \qquad (11.2.4)$$

For a three-dimensional world, the defects of circumferences depend on the plane on which the circles in question are drawn, yet it is possible to define an average curvature by means of the defect from $4\pi r^2$ of the measured area of a sphere of radius r. The result happens to be the following:

$$\text{area} = 4\pi r^2 \left(1 + \frac{1}{9} r^2 R\right), \qquad (11.2.5)$$

where R is the scalar obtained by reducing the curvature tensor twice.

The connection of this idea to the theory of gravitation may be made if we attempt to give a conceptual significance to the sum $R^{12}{}_{12} + R^{23}{}_{23} + R^{13}{}_{13}$, the tensor $G^4{}_4$, which is equal to the 44 component of the stress-energy tensor.

The sum is precisely what we should call the average curvature R of the three-space which is perpendicular to the time axis. Thus we may give an interpretation of the theory of gravitation in words as follows: Consider a small three-dimensional sphere, of given surface area. Its actual radius exceeds the radius calculated by Euclidean geometry ($\sqrt{\text{area}/4\pi}$) by an amount proportional to the amount of matter inside the sphere ($r - \sqrt{\text{area}/4\pi} = G/3c^2\, m_{\text{inside}}$) (one fermi per 4 billion metric tons).

This interpretation serves directly for the 44 component, which is the matter (or energy) density for matter inside the sphere. The other components of the curvature tensor are correctly deduced when we require the same result to hold in any coordinate system regardless of its velocity.

11.3   THE SCHWARZSCHILD METRIC, THE FIELD OUTSIDE A SPHERICAL STAR

In terms of the functions ν and λ, the expressions for the tensor $G^\mu{}_\nu$ are the following.

$$G^4{}_4 = \frac{1}{r}\lambda' e^{-\lambda} - \frac{1}{r^2}\left(e^{-\lambda} - 1\right) = -\frac{1}{r^2}\frac{d}{dr}\left(r\left(e^{-\lambda} - 1\right)\right)$$

$$G^1{}_1 = -\frac{1}{r}\nu' e^{-\lambda} - \frac{1}{r^2}\left(e^{-\lambda} - 1\right)$$

$$G^4{}_1 = \frac{1}{r} \dot{\lambda} e^{-\nu}$$

$$G^1{}_4 = -\frac{1}{r} \dot{\lambda} e^{-\lambda}$$

$$G^2{}_2 = \frac{e^{-\lambda}}{2r}(\lambda' - \nu') - \frac{e^{-\lambda}}{4}\left(2\nu'' + (\nu')^2 - \lambda'\nu'\right) + \frac{e^{-\nu}}{4}\left(2\ddot{\lambda} + (\dot{\lambda})^2 - \dot{\lambda}\dot{\nu}\right)$$

$$(11.3.1)$$

It is only the expression for $G^2{}_2$ which is cumbersome, but it happens that one seldom needs to use it explicitly. The important point is that the divergence of the tensor should be zero. If we have the other components, requiring a zero divergence often avoids the explicit use of $G^2{}_2$. The following exercises may be suggested at this point.

1. Prove that if there is no matter inside of a sphere of radius b, and the matter distribution outside of b is spherically symmetric, the space inside the sphere is flat, with $g_{\mu\nu} = \eta_{\mu\nu}$.

2. Prove that if the stress-energy tensor $T^{\mu\nu}$ is known everywhere inside of a sphere of radius b, whatever is outside b makes no difference to the physics inside b. (The outside is assumed spherically symmetric.)

The solution outside of a spherical mass distribution is obtained by setting $T_{\mu\nu} = 0 = G_{\mu\nu}$, and solving the resulting differential equations.

We begin by noting that $G^4{}_4$ depends only on λ. Since the element $G^4{}_4$ is zero, we have that

$$r\left(e^{-\lambda} - 1\right) = \text{constant} = -2m. \qquad (11.3.2)$$

The factor 2 is conventional, so that the constant m is the total mass of the star times Newton's gravitational constant. If there are no singularities inside the radius a which contains all the mass, the constant must be

$$\int_0^a dr\, r^2\, G^4{}_4 = 2m. \qquad (11.3.3)$$

We are sure that there is no time dependence because

$$G^1{}_4 = 0 = -\frac{1}{r} \dot{\lambda} e^{-\lambda},$$

so that λ is independent of the time in general. The last task is to get an expression for ν. We do this by equating $G^1{}_1$ and $G^4{}_4$, since they are both zero. The conclusion is that

$$\nu' = -\lambda', \qquad (11.3.4)$$

which can happen only if ν is of the following form

$$\nu = -\lambda + f(t), \tag{11.3.5}$$

where f is some arbitrary function of the time. However, since the function ν appears in the coefficient of $(dt)^2$ in the metric as follows:

$$e^\nu (dt)^2 = e^{-\lambda} e^{f(t)} (dt)^2$$

We may eliminate the factor $\exp(f(t))$ by a change of scale of the time coordinate. The other elements of the metric tensor are unaltered by this change, since only the function $\lambda(r)$ is involved. The result is known as the Schwarzschild metric,

$$(ds)^2 = \left(1 - \frac{2m}{r}\right)(dt)^2 - \frac{(dr)^2}{1 - 2m/r} - r^2(\sin^2\theta(d\phi)^2 + (d\theta)^2). \tag{11.3.6}$$

It is interesting that this metric is independent of the time, although at no time have we specified that our search was for a static solution. The time-independence of the Schwarzschild metric follows from the assumption of spherical symmetry, and that we are placed in a region of zero stress density.

For the case of a real star such as the sun, there is no true spherical symmetry because of the rotation and because of the bulge at the equator. However, these produce only small deviations from the spherical case. If there is light streaming outward from the star, there will be another correction because the energy density will not be zero in the space outside. Nevertheless, the Schwarzschild solution is sufficiently close to the situation of the sun, that the precession of the perihelion of Mercury is given correctly to within the observational errors.

11.4 THE SCHWARZSCHILD SINGULARITY

The metric eq.(11.3.6) has a singularity at $r = 2m$. In order to find out whether this is a physically troublesome or meaningful singularity, we must see whether this corresponds to a physical value of the measured radius from the origin of the coordinates (which is not the same as our coordinate r!)

$$R = f(r) \tag{11.4.1}$$

We obtain the answer by considering the metric in another way. We might have supposed that the correct description of a spherically symmetric metric would have been

$$(ds)^2 = H(R)(dt)^2 - F(R)\Big((dx)^2 + (dy)^2 + (dz)^2\Big), \tag{11.4.2}$$

where
$$R^2 = x^2 + y^2 + z^2.$$

The Schwarzschild metric reduces to this form by the substitution

$$r = R + \frac{m^2}{4R} + m, \tag{11.4.3}$$

which results in the expression

$$(ds)^2 = \frac{(1 - m/2R)^2}{(1 + m/2R)^2} (dt)^2 - (1 + m/2R)^4 \Big((dx)^2 + (dy)^2 + (dz)^2 \Big).$$
$$\tag{11.4.4}$$

The singularity in the proper time intervals has disappeared. We see that it was due to a peculiarity in the definition of the radial coordinate r. Nevertheless, the metric eq.(11.4.4) singles out a particular value of the radius, $R = m/2$, as a location at which the coefficient of $(dt)^2$ vanishes. We ought still to investigate what happens to physical processes at this point.

The results need not have any immediate physically observable consequence. When we put in numbers corresponding to the mass of the sun, we find that such a critical radius would exist only if the mass of the sun were concentrated inside a sphere of only 1.5 kilometers radius. However, even though the situation will apparently not occur in the solar system, it is wise to investigate this critical radius as a feature of the theory.

The physical interpretation of this point concerns the rate at which processes occurring near the sun would appear to observers farther out. We have previously calculated how the light from regions of lower gravitational potential is shifted down in frequency, so all things look redder. The radius $R = m/2$ corresponds to a potential so low that the light would not be energetic enough to leave the star, so that no light whatsoever would reach an observer standing at a very great distance away.

We may see whether anything catastrophic occurs to the geometry of space at this point, by computing the curvatures explicitly. They are found to be as follows.

$$\begin{aligned}
R^{12}{}_{12} &= R^{13}{}_{13} = -m/r^3 \\
R^{23}{}_{23} &= 2m/r^3 \\
R^{41}{}_{41} &= 2m/r^3 \\
R^{42}{}_{42} &= R^{43}{}_{43} = -m/r^3
\end{aligned} \tag{11.4.5}$$

We see that the space at the critical point is smooth. The *"singularity" may be nothing but a result of a particular way of choosing the coordinates.* In our example of the bug crawling on the surface of the sphere, there was a singularity in the description of the space upon crossing the equator. But, of course, in a physical sense the space is just as smooth there as anywhere else on a real sphere.

The result we have just obtained, that the curvatures are proportional to $1/r^3$, appears so simple that we might look about for a simple way of getting it. I always feel that a simple result ought to be obtained in a simple way. We shall therefore consider a geometrical argument that reproduces the $1/r^3$ dependence for the case at hand. We shall need once again the concept of mean curvatures, in the three-dimensional space defined by taking the four-dimensional space at an instant of time. In this subspace, the curvatures are analogous to stresses. As for the stresses (or angular momenta) the curvatures define something in a plane, and we may label the components either by the pair of indices which defines a plane, or by the index of the axis perpendicular to the plane. Thus we have an identification

$$R^{12}{}_{12} \to P^3{}_3, \qquad R^{13}{}_{13} \to P^2{}_2, \quad \text{etc.} \qquad (11.4.6)$$

Next we shall demonstrate that the requirement that the divergence of these "stresses" should vanish is equivalent to the Bianchi identity;

$$P^3{}_{3;3} + P^3{}_{1;1} + P^3{}_{2;2} = 0, \qquad (11.4.7)$$

which means that in the space in question the "stress" leads to a zero net force. Superscripts refer to the plane in which the curvatures are being considered.

In dealing with stresses, the trace of the stress tensor is the pressure. In our case, the trace of our stress is the mean curvature, which in turn is the density of matter. We obtain the dependence on $1/r^3$ by demanding in polar coordinates that there should be physical balance in a place of zero pressure. We must be careful to define the areas across which our stresses are acting since they must be physical areas measured along the geodesics. We define the distance along an arc at constant radius r to be $\theta_0 r$, where θ_0 is a small angle. The measurement of θ_0 is well defined, since we go once around and call the total angle 2π. If the radial stress is called T and the hoop stress is called S (see Figure 11.1), we have for a volume element $r^2\, dr\, \theta(\sin(\theta)\, d\phi)$ for which $\theta_0 = \sin(\theta)\, d\phi$ that the forces are unbalanced unless

$$d(Tr^2\theta_0^2) = 2S\, r\, \theta_0^2\, dr.$$

If T is allowed to depend only on r, we obtain the following differential equation connecting T and R,

$$\frac{dT}{dr} = \frac{2S}{r} - \frac{2T}{r}, \qquad (11.4.8)$$

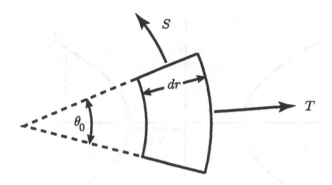

Figure 11.1

which holds in general. We may now specialize to the situation in empty space, in which the trace of the tensor is zero.

$$\text{Trace} = T + 2S = 0, \qquad T = -2S \qquad (11.4.9)$$

The differential equation in this case is

$$\frac{dT}{dr} = -\frac{3T}{r}, \qquad (11.4.10)$$

which has a solution $T = 1/r^3$.

In this way we see how it is the Bianchi identity which implies that the curvatures are proportional to $1/r^3$ everywhere. The connection of the function $\exp(-\lambda)$ to the quantity T may be made by similar simple considerations, which lead to the conclusion that $\exp(-\lambda)$ differs from 1 by an amount inversely as r, eq(11.4.5).

11.5   SPECULATIONS ON THE WORMHOLE CONCEPT

The considerations of the preceding section have shown us how the effect of a spherical mass distribution of sufficiently small size is to produce a curvature proportional to m/r^3 everywhere. A two-dimensional analogue of such a situation would be experienced by a bug crawling on a surface having a "whirlpool" shape. Let us imagine a figure of revolution about a z axis which intersects the xy plane at right angles as illustrated in Figure 11.2. Such a surface may represent our space at a given time ($dt = 0$) and at a definite value of the azimuthal angle, say $\phi = 0$. If the equation of the surface is given by $z(r)$, an arc length at constant angle θ is given by

$$(ds)^2 = (dr)^2 + \left(\frac{\partial z}{\partial r}\right)^2 (dr)^2 = \left[1 + \left(\frac{\partial z}{\partial r}\right)^2\right] (dr)^2. \qquad (11.5.1)$$

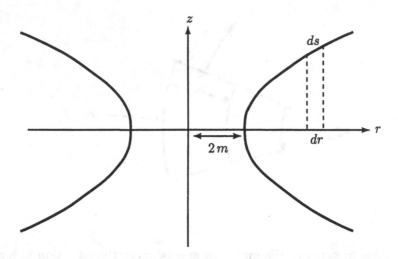

Figure 11.2

We may set the factor of $(dr)^2$ equal to the Schwarzschild value,

$$1 + \left(\frac{\partial z}{\partial r}\right)^2 = \frac{1}{1 - 2m/r}, \qquad (11.5.2)$$

and solve for $z(r)$. The result is easily obtained to be

$$z^2(r) = 8m(r - 2m). \qquad (11.5.3)$$

In other words, the space is parabolic, with the throat of the paraboloid at a distance $r = 2m$ from the origin.

There are some highly suggestive aspects of this result. In the region $r > 2m$, the space is precisely what we would describe as being the result of a mass m about the origin. As we approach the origin, however, we never reach distances $r < 2m$ but cross into another space which is the double of the one we started in. This leads to the idea (developed particularly by Wheeler) that the effects of what we have called a "mass" might be nothing but a peculiarity of the topology of the space in which we are, and that there are "truly" no gravity sources anywhere.

It is exciting to speculate that all the particles with mass must have such throats at a radius $2m$ associated with them—and that it might be that the elementary particles are nothing but regions of space in which we may cross over into the other space by crawling through the hole. The holes are called "wormholes" by J. A. Wheeler. If the particles are charged, the electric field lines might be continuous along the surface, crowding in on one side of the wormhole and crowding out at the other, so that the double space might be associated with the existence of particle-antiparticle pairs.

It has not as yet been possible to obtain a consistent quantitative picture of elementary particles as such wormholes. Neither is there likely to be any experimental observation of any effect due to the wormholes. We do not know of any stars which approach the mass concentration needed for the critical radius to be close to the actual radius. If there were a star smaller than its critical radius, we would not see it because light could not leave the surface, so it is still conceivable that such objects exist. All the known elementary particles have a known structure much larger than the diameter of the associated wormhole. For example, for a neutron we have $r = 2m \approx 10^{-33}$ cm, some 10^{20} times smaller than the known dimensions of the neutron. It would take a particle of 10^{-5} grams for the wormhole diameter to be of the same size as the Compton wavelength $\hbar/mc$.

11.6   PROBLEMS FOR THEORETICAL INVESTIGATIONS OF THE WORMHOLES

There are various questions that may be asked, and these form the basis of theoretical investigations that ought to be very worthwhile. First of all, we may ask, given our present knowledge of the behavior of matter, is it possible that a sufficiently large amount of mass concentrates and collapses into a region smaller than the critical radius? Suppose that we started out with dust spread out nearly uniformly over an enormous region of space. There would start a gravitational collapse, things would warm up, there would be first chemical, then nuclear reactions. As the mass condensed, there would be a point at which the electrons would exert a tremendous pressure resisting collapse, since they cannot be squeezed together closer than the exclusion principle allows. But for sufficiently large masses, the gravitational attractions are big enough to squeeze out the electrons and let the nucleons keep condensing.

The process of this condensation has not yet been theoretically studied in detail. It seems to me that before we shall know anything about our wormholes we need to work out problems of the classical theory of gravity with very large masses. If the collapse into a sphere smaller than the critical radius is possible, we shall never see it from the outside, because the light would get redder and redder and then infrared and then radio and finally D.C. But there would be physical sense in asking what it would be like to be part of the collapsing mass.

Let us see how we would go about describing physical processes occurring in a reference frame moving with the flowing matter. The equation of state would involve the pressure p and the matter density ρ. In static matter, we would have

$$T^1{}_1 = T^2{}_2 = T^3{}_3 = -p, \qquad T^4{}_4 = \rho. \qquad (11.6.1)$$

What does this tensor look like for a moving piece of matter? In terms of the quantities

$$u = \frac{v}{\sqrt{1 - v^2/c^2}}, \qquad w = \frac{1}{\sqrt{1 - v^2/c^2}}, \qquad \rho_0 = \rho + p, \qquad (11.6.2)$$

we find

$$\begin{aligned}
T^4{}_4 &= \rho_0 w^2 + p, \\
T^1{}_1 &= -\rho_0 u^2 - p, \\
T^2{}_2 &= T^3{}_3 = -p, \\
T^4{}_1 &= -\rho_0 u w.
\end{aligned} \qquad (11.6.3)$$

To solve the dynamical problem of the mass condensing from dust, we might proceed as follows. First, we assume that the situation is to be described by means of the functions λ and ν of the radius and the time. We assume that the state of the matter is describable by the matter density ρ and the pressure p. We need also an equation of state to relate p and ρ;

$$p = f(\rho). \qquad (11.6.4)$$

As a first attempt, we may see what happens if p and ρ are connected adiabatically. Later, we may see what happens as we allow cooling by radiation of light, heating by nuclear reactions, etc. What we want is to obtain an answer for the radial velocity of a piece of matter, $u(r, t)$.

The problem that we would get including heat flows and opacities and what-not might be very complicated, but the answer ought to be obtainable as the solution of a set of connected partial differential equations. The hope is that this set would be consistent and that the dependences could be disentangled so the equations could be solved in sequence according to some procedure. It would be too much to hope for solutions in closed form. If the differential equations were disentangled, however, we might hope that machines could provide us with numerical solutions of the set.

12.1 PROBLEMS OF COSMOLOGY

In the last lecture we have briefly outlined one of the problems of the classical theory of gravitation, that of a spherically symmetric mass distribution, which is an idealized model for a star. The second problem we tackle by classical gravitation is that of cosmology, the "science of the universe." All other problems in the theory of gravitation we shall attack first by the quantum theory; in order to obtain the classical consequences on macroscopic objects we shall take the classical limits of our quantum answers.

It is very difficult to state precisely what cosmology is. In general it deals with all that can be known about what happens, if the scale is gigantic, sufficiently large that even galaxies may be treated as being infinitesimal in size. It may also deal with the question of where visible matter came from, starting from a given initial hypothesis such as "In the beginning, all was hydrogen."

One aspect of cosmology deals with the actual geography of matter; the question of importance is where the matter is and what it is doing there. The relevant observations are of how many galaxies are East or West and in what direction they are moving. We believe that the motion of the galaxies is governed solely by gravitation, so that, once we have seen or measured the distribution of matter and velocity, it is a simple physical problem to predict what happens next. Cosmological problems of a different nature appear when we go to such an enormous scale that the detailed structure should disappear. The problem of what happens next can in principle be solved given any starting conditions whatsoever. When all the detailed motion is averaged, however, we may ask whether the universe is static or evolving, whether stable or unstable, whether finite or infinite. One intriguing suggestion is that the universe has a structure analogous to that of a spherical surface. If we move in any direction on such a surface, we never meet a boundary or end, yet the surface is bounded and finite. It might be that our three-dimensional space is such a thing, a tridimensional surface of a four sphere. The arrangement and distribution of galaxies in the world that we see would then be something analogous to a distribution of spots on a spherical ball.

It turns out that the theory of gravitation alone does not yield an answer which limits the possible averaged distributions over the sphere, nor does it confirm or deny that the universe is bounded or infinite as a hyperbolic paraboloid. Thus the problems of cosmology are always tied up with some fundamental assumptions. The most reliable way to examine the validity of these assumptions is to deduce some consequences and then compare the results with observations.

The observations on the geography of matter which is very far away are very difficult and rather uncertain even with the latest techniques. It must also be borne in mind that a large portion of the sky is blocked to observation by our own galaxy, which has so much dust that we cannot make observations in directions along the galactic plane. In spite of these difficulties and possible qualifications, the evidence to date suggests that the universe is uniform throughout, with galaxies randomly here and there, some bigger, some smaller, but so that any given large region is much like any other large region. To give a two-dimensional analogue, we say it is as though a car ran through a puddle and splattered mud spots randomly on a wall, and we are sitting on one mud spot looking out at all the other spots.

The velocity distribution is very interesting, if we compare the velocity of galaxies with their apparent distance from us. Let us skip over the difficulty of distance measurements, which is considerable. The astronomers have gotten some distances by hook and crook, by assuming statistical distributions of brightness for example, and they are willing to quote them with some uncertainty or other, which is getting smaller all the time. At the same time, we have a measure of the velocity of galaxies by the Doppler shift of spectral frequencies. The results are consistent in that they show in the light from a given object a shift toward lower (redder) frequencies, which is proportional to the distance that the object appears to be from us.

In Lecture 1 we discussed a simple model which serves to interpret the facts. If all the masses in the universe are fragments from an explosion which occurred a time T ago, and we assume that gravitational forces are weak, then we expect that a fragment which has been moving with a velocity V relative to the center is to be found now at a distance $R = VT$ away. This equality holds no matter what magnitude the velocity V may be, so for all fragments we expect $(R/V) = T$, a universal constant. The observations are consistent with T constant and having a value in the range $(10 - 13) \times 10^9$ years. The uncertainties come not from the velocity measurements, but from the distance measurements. The farthest objects which have been seen are receding from us at a speed $(v/c) = 0.48$. This red shift is one of the key observations which tells us something about the universe.

Other observations concern the distribution of galaxies. Although all visible parts of the sky are remarkably alike, galaxies are not randomly distributed but exist in clumps or clusters. We would say that galaxies were randomly placed if we found that for different regions of the universe of given dimensions the number of galaxies is a constant N with a spread $\pm\sqrt{N}$. On the average, the distances between galaxies are some ten times their diameters. Our nebula has a diameter of 10^5 light-years, so that the average intergalactic distance is 10^6 light-years. The distribution of galaxies inside cubes of edges larger than 10^6 years is not $N \pm \sqrt{N}$. Galaxies are found preferentially in clusters of perhaps 50 galaxies as a typical number. In addition, one finds clusters of clusters. However, no clusters of clusters of clusters are said to exist—this means that if we go to a scale which is large compared to 10^8 light years, the universe seems very nearly to have the "random" distribution of galaxies.

Since the clustering of galaxies and the clustering of clusters is assumed to be due to the gravitational interactions between them, it has been suggested that the absence of clusters larger than some 10^8 light-years in radius is evidence for a cut-off in the gravitational force at a distance of this order of magnitude. We shall not adopt this point of view, because we do not want to modify our theory unless there are phenomena actually contradicting it; the absence of clusters of clusters of clusters does not seem to me to be a contradiction of our theory. We take it to be a measure of the scale of length over which we must average matter densities if we want to treat the universe as being in some sense homogenized.

Is there any variation in this homogenized matter density in regions at a different distance from us? The observations attempt to count the number of galaxies in shells having inner radius R and outer radius $R + \Delta R$. The results suggest that there may be a slight variation of density with radius, making the universe denser in farther regions. However the uncertainties involved are large compared with the fractional variation from a constant density; a theory of the universe predicting or assuming a constant density would not be in disagreement with the present estimates.

12.2   ASSUMPTIONS LEADING TO COSMOLOGICAL MODELS

Since the observations are not sufficiently accurate to clearly suggest definite characteristics, we must rely on our ingenuity and make a grand hypothesis about the structure of the universe. The grand hypothesis that nearly every cosmologist makes is that the universe (on a grand cosmological scale larger than 10^8 light-years) looks the same no matter where one may be in it—although not necessarily simultaneously. That is, at any point in the universe there will be a time, or there was a time, at which the universe looked or will look as the universe looks to us now. This means that provided we shift the time scales to match corresponding times appropriately, the evolution of the universe follows the same path no matter from where we look at it.

The assumption we have just mentioned implies a very strong uniformity in the universe. It is a completely arbitrary hypothesis, as far as I understand it—and of course not at all subject to any kind of observational checking, since we have been and will continue to be confined to a very small region about our galaxy, and the time development of the universe follows a "cosmological" scale a billion times longer than our lifetime. I suspect that the assumption of uniformity of the universe reflects a prejudice born of a sequence of overthrows of geocentric ideas. When men admitted the earth was not the center of the universe, they clung for a while to a heliocentric universe, only to find that the sun was an ordinary star much like any other star, occupying an ordinary (not central!) place within a galaxy which is not an extraordinary galaxy but one just like many many others. Thus, it is assumed that our place in the universe should be just like any other place in the universe, as an extension of the sequence I have described. It would be embarrassing to find, after stating that we live in an ordinary planet about an ordinary star in an ordinary galaxy, that our place in the universe is extraordinary, either being the center or the place of smallest density or the place of greatest density, and so forth. To avoid this embarrassment we cling to the hypothesis of uniformity.

Yet we must not accept such a hypothesis without recognizing it for what it is. An analogy will illustrate my viewpoint. If we parachute out from a plane flying at random over the earth and land in a clump of birch trees about this spot we might argue, we landed at random in no particular spot—there being nothing unique about this spot we conclude that the earth is covered with birch trees *everywhere*! The conclusion would be false regardless of the perfect randomness of the place where we might land. Yet perhaps we are doing the same thing in constructing our fundamental assumption of cosmology.

We shall mention but three theories of the universe. There is a theory due to Milne [Miln 34] which disregards the gravitational forces altogether; it is a good theory if the average density of the universe is sufficiently small. There is a theory originally due to Einstein and pursued by others, which arose out of the assumption that the universe is static, rather than dynamic. The assumption predated Hubble's observation of a red shift proportional to distance. The static universe could not be maintained without adding a term to the stress tensor in Einstein's equations, as follows.

$$R^\mu{}_\nu - \frac{1}{2} g^\mu{}_\nu R = KT^\mu{}_\nu + \Lambda g^\mu{}_\nu \qquad (12.2.1)$$

The multiplying constant Λ is known as the "cosmological constant." We have discussed the possible appearance of these terms, which come from a piece of the action which is

$$\Lambda \int d\tau \sqrt{-g}. \qquad (12.2.2)$$

Had Einstein decided that such a term could not be present in his equations, he would have predicted an evolving universe such as Hubble observed. After Hubble's discovery Einstein was no longer interested in such a cosmology, which has fallen into discredit although many people still try to work out theories using various values for the cosmological constant. We shall only consider theories for which $\Lambda = 0$.

An ingenious theory is due to Hoyle [Hoyl 48], who assumes not only that the universe evolves everywhere along a similar path, but that the universe is actually in a steady-state—it looks the same everywhere and at all times. In order to maintain this universe, in which stars and planets are continuously being created from some cosmic dust, there must be a continuous creation of matter everywhere in the universe, so that as the condensed galaxies recede from each other the average density remains constant. No mechanism is specified for the creation of such matter, and the theory neglects to consider the details of conservation of energy; for example no mechanism is described by which one can understand what the state or the velocity of the matter is at its creation. Although in ordinary circumstances a physicist would rebel against a theory which so cavalierly ignores our cherished laws about the conservation of matter and energy, one must remember that here we are not dealing with an ordinary problem but with a cosmological problem. The other cosmological theories sweep creation of matter under the rug by simply assuming a time at which the matter was already there, and speaking only of what comes after. Again, no mechanism is given for the creation, so that the steady-state theory can hardly be accused of being unreasonable on this account. It should also be borne in mind that the universe is so vast that the rate of creation could well be extremely small, much too small to have been observed directly. If only one atom of hydrogen per cubic mile of space were created every year, that would sustain the universe in its steady-state.

We shall first discuss a theory with $\Lambda = 0$ which does not assume that the universe looks the same at all times, but which assumes that the universe follows an identical development at all places. If we arrange the time scales referred to different origins so that the corresponding stages are denoted by the same value of the coordinate t, the Robertson-Walker metric which defines the geometry is assumed to be

$$(ds)^2 = (dt)^2 - \frac{R^2(t)}{(1 + kr^2/4)^2}\left[(dr)^2 + r^2\left(\sin^2\theta\,(d\phi)^2 + (d\theta)^2\right)\right].$$

$$(12.2.3)$$

Let us state some of the simple properties of this metric. If we stay at the same place, then $ds = dt$, no matter where we are. If we look at the universe at any given time $dt = 0$, the three-space at a given time is spherically symmetric but may have a certain curvature. The idea of the uniformity of space requires this spherical symmetry, since a spherical surface is the only kind of surface which looks the same no matter where one may be on it. Thus, we write a metric which corresponds to a three-dimensional surface of constant curvature, which is isotropic from every point. For $k > 0$ the metric corresponds to a three-sphere, for $k = 0$ we have flat space, and for $k < 0$ we have negative curvature and an unbounded universe.

Let us see how we would describe a three-surface which is spherical. We use mathematics analogous to that of the two-dimensional spherical surface, which is described by two angles, θ and ϕ; the surface is at a constant radius b, and the angles are defined so that

$$z = b\cos\theta,$$
$$x = b\sin\theta\cos\phi, \qquad x^2 + y^2 + z^2 = b^2 \qquad (12.2.4)$$
$$y = b\sin\theta\sin\phi.$$

In four dimensions all we do is invent a third angle ξ such that

$$w = a\cos\xi,$$
$$z = a\sin\xi\cos\theta,$$
$$x = a\sin\xi\sin\theta\cos\phi, \qquad (12.2.5)$$
$$y = a\sin\xi\sin\theta\sin\phi.$$

Using this angle ξ, the metric on the three-surface $dt = 0$ is proportional to the square of the radius and to

$$(d\xi)^2 + \sin^2 \xi \left((d\theta)^2 + \sin^2 \theta \, (d\phi)^2\right). \qquad (12.2.6)$$

To make the transition to the metric eq.(12.2.3) in terms of a radial coordinate, we simply introduce a transformation such that, $(r^2 \neq x^2 + y^2 + z^2)$,

$$\frac{dr}{r} = \frac{d\xi}{\sin \xi}, \qquad (12.2.7)$$

which results in an expression for $\cos \xi$.

$$\cos \xi = \left[\frac{1 - \frac{kr^2}{4}}{1 + \frac{kr^2}{4}} \right] \qquad (12.2.8)$$

When we compare expressions, we find that the metric eq.(12.2.3) truly represents a three-surface which is indeed spherical. $R(t)$ is the conversion factor between the coordinate differentials and arc lengths, which varies with time; the metric is not static, in general.

12.3   THE INTERPRETATION OF THE COSMOLOGICAL METRIC

The first question that we should examine is that of the dynamics of objects in this metric. Do objects at rest remain at rest? For such objects only u^t is nonzero and the equation of motion reduces to

$$\frac{du^\alpha}{ds} = -\Gamma^\alpha_{tt} \, u^t \, u^t.$$

Since $\Gamma^\alpha_{tt} = 0$, this system of coordinates can be realized by a collection of massive particles.

By use of the metric eq.(12.2.3) we should be able to make some predictions on observable quantities, in terms of $R(t)$. It may then be possible to construct a model for the universe which specifies $R(t)$. For example, Hoyle's steady-state universe corresponds to an $R(t)$ which is exponential in time. Milne's universe corresponds to $R(t)$ which is simply proportional to the time t.

Let us briefly state some of the properties of Milne's universe, in order to understand just how the universe looks the same at corresponding times in different locations. We have said that this model corresponds to zero gravity. It is, for example, possible to rearrange coordinates so the resulting metric is flat. However, as we have written it, our three-surface appears to have a curvature. We can understand the source of

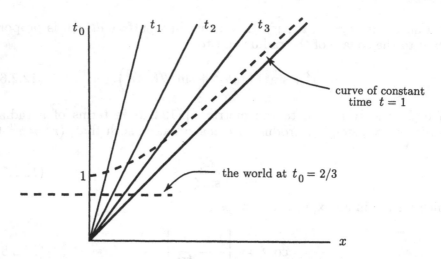

Figure 12.1

this curvature by considering a set of clocks with identical tickers which diverge at different speeds from a given origin at a given time, $t = 0$. If we make a Minkowski diagram, the world-lines of the various clocks represent the time axes of various coordinate systems. Thus, the surfaces of equivalent times are curved (Figure 12.1). The three-surface "at a given time" corresponds to such a hyperbolic surface, and therefore has some curvature. However, the four-dimensional space is just as flat as before.

A more nearly operational viewpoint of such a curvature concerns the measurement of a radius and a circumference. If we measure them in a subspace "now" we find the circumference is $2\pi R$. The measurements at constant t correspond to measurements made as though the local nebulae, streaming outward with a velocity proportional to the distance r, were at rest. Thus, the radius is measured by rods moving *along* their lengths. The circumference is measured by rods which are moving perpendicular to their lengths, and hence the observation of an effective curvature.

Now that we have clarified the significance of the three-surfaces of constant t, let us see what we can deduce about astronomical observations. For example, let us deduce the detected change in apparent frequency of a signal which is emitted somewhere else, in terms of the function $R(t)$. We must ask what is the interval of reception if we specify a time of emission and the coordinate r. The path of a light signal is given by setting $(ds)^2 = 0$. If the light travels along a radius, then we have

$$(ds)^2 = 0 = (dt)^2 - \frac{R^2(t)}{(1 + kr^2/4)^2} (dr)^2,$$

which implies that

$$\frac{dt}{R(t)} = \frac{dr}{(1 + kr^2/4)}. \tag{12.3.1}$$

If we are discussing a clock at constant coordinate r, then dt is the proper time interval; the number of ticks of a clock is simply $\int dt$. In order to compare frequencies we must calculate the ratio of the length of the emission time to the reception time. If we send a signal which starts at a time t, and ends at a time $t + \Delta t$, it will be received starting at some time t_0 and reception will end at a time $t_0 + \Delta t_0$. Integrating along the light path eq.(12.3.1) of the starting pulse and end pulse, we find

$$\int_t^{t_0} \frac{dt'}{R(t')} = \int_0^r \frac{dr}{1 + kr^2/4} = \int_{t+\Delta t}^{t_0+\Delta t_0} \frac{dt'}{R(t')}, \tag{12.3.2}$$

because the coordinate r is a constant for a given nebula. Whatever the length of the interval $t_0 - t$ might be, if Δt and Δt_0 are both short compared to the variations in $R(t)$, eq.(12.3.2) implies that

$$\frac{\Delta t}{R(t)} = \frac{\Delta t_0}{R(t_0)}. \tag{12.3.3}$$

Thus, the comparison of frequencies is determined by the function $R(t)$; the result is

$$w_{received} = \frac{R(t)}{R(t_0)} \cdot w, \qquad w = \text{natural frequency}. \tag{12.3.4}$$

The function $R(t)$ is evidently a scale factor for the universe. If $R(t)$ is monotonically increasing with t, that is, corresponding to an expanding universe, all received frequencies are red-shifted. The amount of the shift will be approximately proportional to $(t_0 - t)$ if this interval is small compared to the variation of R.

Before we may connect this red shift with Hubble's red shift, we must devise a scheme for the assignments of distances corresponding to different values of the coordinate r.

12.4   THE MEASUREMENTS OF COSMOLOGICAL DISTANCES

Let us imagine that we attempt to measure the distance to a faraway nebula by looking at its apparent angular diameter. We suppose that the nebula has a diameter L; then if we observe an angular spread $\Delta\theta$, the distance to the nebula D_0 should be given by

$$\Delta\theta \cdot D_0 = L, \quad \text{i.e.,} \quad D_0 = \frac{L}{\Delta\theta}. \tag{12.4.1}$$

The length L is supposed to be the length at equal times, $dt = 0$. We assume that ϕ remains constant and that $\Delta\theta$ may be treated as infinitesimal. The proper time interval is space-like:

$$-(ds)^2 = \frac{R^2(t_0)}{(1 + kr^2/4)^2}\, r^2\, (\Delta\theta)^2 = L^2, \qquad (12.4.2)$$

so that

$$D_0 = \frac{R(t_0)r}{(1 + kr^2/4)}. \qquad (12.4.3)$$

Over a range of distances much smaller than $1/\sqrt{k}$, r will correspond to the distance measured in this way, with a scale factor $R(t)$. We see also that using distances measured in this way, then the red shift of faraway galaxies should be proportional to their distance from us (Hubble's law), as a first-order result.

Another common method of estimating distances makes use of the apparent brightness of galaxies. It is assumed that galaxies have a constant average "standard" brightness, emitting a given number of photons every second. The method is analogous to estimating the distance of a standard candle by saying $D^2 =$ (Standard Intensity/Apparent Intensity), since intensity follows an inverse-square law. For the present problem, the apparent solid angle is (L^2/D^2), where L is the diameter of a galaxy; we must include also a time dilation factor, since the N photons emitted in our direction in an interval Δt will be observed in an interval Δt_0, related to Δt by eq.(12.3.3). If we compare intensities, we must include a factor which expresses the decrease in energy of photons because of the frequency shift eq.(12.3.4). The net result is the following for the relation between a distance D and r.

$$D = \frac{R^2(t_0)r}{R(t)(1 + kr^2/4)} \qquad (12.4.4)$$

This differs from the result eq.(12.4.3) by a factor $R(t_0)/R(t)$, so that the numbers D_0 and D do not coincide. It is nevertheless possible to hook all this together and solve for $R(t)$ in the ideal case; these considerations are the drive for the observations made with the 200-inch telescope at Mt. Palomar; it spends a great deal of its time looking at galaxies, measuring diameters, intensities, and red shifts, in a search for better evidence on the nature of the function $R(t)$ if the present model were correct.

If we ask for the number of nebulae that ought to be in a shell of thickness $d\rho$ at a distance ρ from us, we get of course different expressions depending on whether ρ means the D_0-type of distance or the D-type. Nevertheless the answer turns out to be,

$$\text{No. between } (\rho) \text{ and } (\rho + d\rho) = \frac{dr \, r^2 R^3(t)}{(1 + kr^2/4)^3} \cdot (\text{constant}), \quad (12.4.5)$$

assuming nebulae have the same average mass at all radii.

It must be emphasized that all of these methods of investigating the structure of the universe have built-in assumptions which may very well be incorrect. The determination of distance from brightness assumes no significant change of the brightness of nebulae with age. Some astronomers have attempted to calculate elaborate corrections for the assumed evolution of stars, but in truth we don't know just how intensities evolve in old nebulae. Should we rather measure diameters? Not only are diameters hard to measure for faraway nebulae, but also it is not known whether the apparent diameters increase or decrease with age. There are further difficulties in that when galaxies become very faint it is nearly impossible to be very sure of how many we are losing because of their dimness. These difficulties do not affect the results provided we assume Hoyle's universe; his is the only completely detailed cosmology, and it specifies implicitly that nebulae on the average should *be* the same, since the universe is in a steady-state.

12.5   ON THE CHARACTERISTICS OF A BOUNDED OR OPEN UNIVERSE

The detailed dynamics of the universe (called Friedmann models when $\Lambda = 0$ and Lemaître models when not) may be studied in terms of the elements of the stress-energy tensor. If we calculate these from the curvature tensor deduced from the metric eq.(12.2.3) we obtain for the 44 component

$$T^4{}_4 = 3 \left(\frac{k + \dot{R}^2}{R^2} \right) = 8\pi G\rho + \Lambda, \quad (12.5.1)$$

where ρ is the average matter density. The other diagonal elements are

$$T^1{}_1 = \frac{2\ddot{R}}{R} + \frac{k + \dot{R}^2}{R^2} = -8\pi Gp + \Lambda. \quad (12.5.2)$$

In this equation, p represents the averaged pressure. It includes *all* pressures, the ordinary gas pressure plus the light pressure and any pressure due to any other process whatsoever. For our discussion, we shall assume that the gas pressure is so much larger than any other that all others may be neglected. But this pressure is very small, since it is of the order of $(1/2\ \rho v^2)$, where (v/c) is an average velocity which is quite small, and we shall presume it plays no role in determining the dynamics of the universe. We obtain the dynamics directly from the density ρ by requiring that T must satisfy a divergence condition. What results is the following relation between p and ρ.

$$\frac{d}{dt}(\rho R^3) = -3pR^2\frac{dR}{dt} \tag{12.5.3}$$

This equation is just

$$T^\mu{}_{4;\mu} = 0.$$

The other independent covariant divergence,

$$T^\mu{}_{1;\mu} = 0$$

gives $p_{,r} = 0$, which is to be expected, since the universe is isotropic. The result has very simple structure, and it has an evident classical meaning if we call R the radius of the universe. ρR^3 is proportional to the total mass, that is, the energy content, of a uniform density ρ of dimension R. The term on the right is the rate of doing work, since it is a pressure times a volume. The equation has the same structure if instead of the whole universe we take a smaller region, of dimension a proportional to R. In this case,

$$\frac{d}{dt}(\rho a^3) = -3pa^2\frac{da}{dt}. \tag{12.5.4}$$

If $p = 0$, then the amount of matter inside a sphere does not change;

$$\frac{4\pi}{3}\rho R^3 = M \tag{12.5.5}$$

is a constant. We can solve the equations to get

$$(k + \dot{R}^2) = 2G\frac{M}{R}. \tag{12.5.6}$$

This differential equation can be solved to find $R(t)$. The behavior of the possible solutions is easy to understand, still within the framework of Newtonian mechanics. What can happen may be easily considered in terms of what can happen to a shell of thickness da outside of a spherical charge distribution of mass m (see Figure 12.2). It can be considered as being in free fall in the field of the mass inside, which is constant, and it will follow a falling-body equation. The energy conservation would tell us in Newtonian mechanics that

$$-\frac{Gm}{a} + \frac{\dot{a}^2}{2} = \text{constant} = \text{Energy/Mass}. \tag{12.5.7}$$

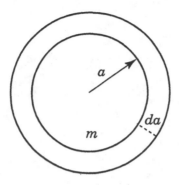

Figure 12.2

Depending on the magnitude of this energy, three types of solutions are possible.

1.   If the energy is positive, the shell keeps expanding forever, and keeps expanding at infinite times.

2.   If the energy is zero, it expands asymptotically toward a static universe of infinite dilution.

3.   If the energy is negative, the motion is bounded and cyclic.

These solutions of the Newtonian problem correspond to the possible types of universe; 1) corresponds to an open universe of negative curvature, 3) corresponds to a closed finite universe of positive curvature.

Why are these Newtonian solutions good enough to characterize our answers? It is because in a problem of spherical symmetry, the motion of a finite shell of matter is determined by the mass inside it only. The mass outside produces an equivalent flat space inside. Thus, by examining the motion of a finite shell we get the behavior of the entire universe. Here again we see the power of the assumption of cosmological uniformity.

Figure 13.2

13.1 ON THE ROLE OF THE DENSITY OF THE UNIVERSE IN COSMOLOGY

Now that we have seen how the postulate of uniformity leads to the possibility of a universe which may be either open or closed, we see that one of the most interesting cosmological questions is whether our universe is unbounded and exploding forever, or bounded. We expect to answer the question on the basis of observations. What are the relevant factors? The central question is, are the velocities of the nebulae sufficiently large that they escape, or are they so small that their motion is bound? Let us make some estimates on the basis of Newtonian mechanics, which is sufficiently close to relativistic mechanics for this purpose. If the radial velocity of a shell at radius r is proportional to r, is the kinetic energy sufficient for escape? Because of the spherical symmetry, we consider only the mass inside the shell in writing down the conservation of energy. If we assume a uniform density ρ for the universe, the critical condition for a barely-bound or barely-unbound universe is that

$$\frac{1}{2}v^2 = \frac{GM}{r} \qquad \text{where} \qquad M = \frac{4\pi}{3}\rho r^3. \qquad (13.1.1)$$

We may do this calculation for any value of r whatever. We now set $v = r/T$ where T is Hubble's time, one of the quantities we must observe. The critical condition may then be expressed in terms of the density

$$\rho = \frac{3}{8\pi} \frac{1}{GT^2}.$$ (13.1.2)

If we accept the present value of Hubble's time, $T = 13 \times 10^9$ years, we calculate this critical average density to be $\rho = 1 \times 10^{-29}$ grams/cm^3. We shall know whether the universe is bound or unbound as soon as we have measured the average density to sufficient accuracy to have a valid comparison with the critical value.

The measurements of the density of the universe are unfortunately extremely difficult and extremely uncertain, much harder than measurement of Hubble's constant, which itself may have a sizeable uncertainty. As we have already mentioned, only a few years ago T was thought to be smaller by a factor 2.4. The revised estimate represents a great deal more effort than the original, so that it may be more reliable, yet it is not inconceivable that the present value may yet be off by a similar factor again. As usual, the final result is very sensitive to the value measured for the most difficult cases to measure, namely, those of galaxies which are at the extreme range of our telescopes. The distances are estimated on the basis of the brightness of clusters rather than the brightness of galaxies, and there is no guarantee that these observable clusters are at all typical of their region, nor that they are of an intensity comparable to the clusters nearer to us. Thus, the final estimates of T are contingent on the correctness of a long chain of assumptions, each one contributing significantly to the possible error in the final result. What is the situation as regards estimates of the average density? If we count galaxies and assume they are more or less like ours, the total density of this kind of visible matter amounts to some 10^{-31} g/cm^3. This represents some kind of lower limit on the matter density, since the visible matter must be a fraction of the total. The density of matter in intergalactic regions has been estimated by measuring the intensity of various sharp spectral lines as a function of the distance of the source. Presumably the absorption of such lines is a measure of the number of atoms of the given type in the intervening region. There are a great many assumptions and corrections to be made in converting the data to a numerical estimate, so that the final result is also not expected to be certain except perhaps to an order of magnitude. Measurements on the amount of sodium have been moderately successful, but the key quantity is the amount of hydrogen, which is presumably to be obtained from the absorption of the 21-cm line of Hydrogen. The radio astronomers are working on this; their results to date tend to indicate that the visible matter is indeed a very small fraction of the total. The critical value, $\rho = 1 \times 10^{-29}$ g/cm^3 is always within the range of any estimate; yet the data has enough slop so that if a theory were to require a density as high as 10^{-27} g/cm^3, the observations could not rule it out, the theory could not be disproved on the basis that the density predicted is much too high.

At this point I would like to make a remark on the present state of observations relevant to cosmology. When a physicist reads a paper by a typical astronomer, he finds an unfamiliar style in the treatment of uncertainties and errors. Although the papers reporting the calculations and measurements are very often very careful in listing and discussing the sources of error, and even in estimating the degree of confidence with which one can make certain key assumptions, when the time comes for quoting a number, such as Hubble's time T, one does not find an estimate of the over-all uncertainty, for example in the form of the familiar $\pm \Delta T$ used by physicists. The authors are apparently unwilling to state precisely the odds that their number is correct, although they have pointed out very carefully the many sources of error, and although it is quite clear that the error is a considerable fraction of the number. The evil is that often other cosmologists or astrophysicists take this number without regard to the possible error, treating it as an astronomical observation as accurate as the period of a planet. An example of the evil occurs when dealing with cosmological models, where all models predict a red shift very nearly linear with distance. It is only in the acceleration of faraway galaxies that differences between the models become significant. If this acceleration happens to be expressed as a parameter q, which observations and calculations show to be, for example,

$$\frac{\ddot{R} R}{\dot{R}^2} = q = 1.2, \tag{13.1.3}$$

the number is useless to compare with theory. Even if Hoyle's model should predict $q = -0.5$, he could still be right, because the acceleration might have huge errors, 1.2 ± 5.0 for example. Yet this is precisely where some people go astray, in treating the numbers without errors as though they were extremely precise. Some day, of course, the errors may be much smaller, since a great deal of effort is being made by many astronomers to get increasingly better numbers.

The upshot of all this is that the critical density is just about the best density to use in cosmological problems. It has a great many pleasing properties—for example, it is the density for which the creation of matter at the center of the universe (which is everywhere, according to the principle of cosmological uniformity) costs nothing. If the density of matter is ρ then we know that the space has a positive curvature due directly to the matter density. On the other hand the three-space of equivalent proper times in the evolution of nebulae has a negative curvature; the critical density balances it to make a flat three-space. The critical density also separates the case of a universe with a finite number of nebulae from a universe with an infinite number. With all of these magic properties of this number, it is exciting to speculate that it indeed is the "true" density—yet we must not fool ourselves into thinking that a beautiful result is more reliable simply because of its "beauty," which is in part an artificial result of our assumptions.

13.2   ON THE POSSIBILITY OF A NONUNIFORM AND NONSPHERICAL UNIVERSE

Since all our conclusions have been so heavily based on the postulate of a uniform cosmology, we might review the nature of the evidence. If we examine a region of the universe within 1.3×10^8 light-years from us, we find simply one cluster, the Virgo cluster—in other words, the matter is distributed in a strongly unsymmetric fashion. The lack of symmetry in this region is so large that it can't be explained away. For example, it cannot be due to obscuring by galactic dust, which could affect only that region of the sky lying within a few degrees of the galactic equator. The uniformity must show up in examining regions which are large compared to 10^8 ly, since the number of galaxies in the usual cluster is much too large to be comfortably described as a fluctuation of the density. What is encouraging is that in very faraway regions the sky seems to be populated by very regular and compact clusters of clusters, involving perhaps thousands of nebulae, all swarming about as bees, relative to the center of gravity of the supercluster. The dispersion of the red shifts shows that the velocities relative to the center of mass must be of the order of 1000 km/sec. This dispersion serves as a very useful measure of the mass of the clusters. It is evident that the clusters are formed by gravitational attractions, and that they are entities which have a long life—they are stable bound systems. From this information we deduce the mass, because we know that velocities of 1000 km/sec are insufficient for escape. There are some worries in this correction; for example if we make a similar computation for the mass of the Virgo cluster, we get a mass 30 times smaller than by other means. Yet the faraway superclusters are larger so that the computation may be more reliable.

The very existence of clusters shows that nebulae attract each other with sufficient strength that these systems are held together for times comparable to the age of the universe. It is very interesting to note that almost all galaxies are in clusters; only a very small fraction appear to exist not in a cluster. The conclusion is that nearly all of the visible matter of the universe does not have sufficient kinetic energy to escape from other matter nearby. In view of this fact, it seems to me that it is very unlikely that the average density of the universe is much smaller than the critical density. If the density is much smaller, the formation of the clusters must be ascribed to local fluctuations which made matter denser in some regions. But on a statistical basis it would be very difficult that there should be enough local fluctuations of the right kind so that nearly all matter occurs in clusters. It might be argued that at an earlier time when the density was greater such fluctuations might be easier—but I have not seen a quantitative argument based on this idea. The inescapable conclusion is that most matter gets pulled into clusters because the gravitational energy is of the same order as the kinetic energy of the expansion—this to me suggests that the average density must be very nearly the critical density everywhere.

The preceding guesses about the average density do not lend any support to the hypothesis of uniformity. It could still be that the question as to whether the matter in given regions holds together is purely local, and the situation varies from region to region. The cosmological principle can be studied only by making a detailed comparison of the density of matter and the nonlinearity of the red shift, which represents an acceleration of the galaxies. These are quantities which are in principle independently measurable, yet a definite relation between them is predicted by theories adopting the cosmological principle. If the variation of density with radial distance were to be measured with great accuracy, we might find that maybe matter is too dense in the inner region, in such a way that the cosmological principle cannot hold.

Even if the cosmological principle were wrong, it would be possible for the universe to have a spherical symmetry—a prejudice in favor of the cosmological principle reflects our surprise at finding that we are at what looks like the center of the universe. Let us assume that the visible region is nearly symmetric but that outside there is more matter, distributed in a lopsided fashion. What would be the difference? We would expect a first-order correction to the motion of the distant nebulae to be due to the tidal forces as illustrated in Figure 13.1. (If the visible region were accelerating as a whole we would not be aware of the acceleration!) The result would be a correction to the red shift having a quadrupole character; the red shift should be redder in two polar regions and bluer in a equatorial region.

The preceding discussions show us how little the relativistic theory of gravitation tells us about cosmology. The central problems of cosmology can be solved only when we actually know what the universe "really" looks like, when we have accurate plots of red shifts and densities as a function of distance and position in the galactic sky.

13.3   DISAPPEARING GALAXIES AND ENERGY CONSERVATION

Let us briefly mention some of the other interesting puzzles of cosmology and cosmological models. Is it possible for certain nebulae to vanish from view forever? With our present theory, the answer is no; regardless of

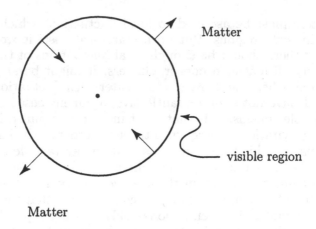

Figure 13.1

whether the density is greater or smaller than critical, no galaxy which is now visible will ever disappear entirely, although it may get redder and redder. It makes no sense to worry about the possibility of galaxies receding from us faster than light, whatever that means since they would never be observable by hypothesis.

In Hoyle's theory, the nebulae far away do vanish; his expression for the recession velocity is such that the volume of a cube defined by eight given galaxies doubles in every time interval of a certain size. The velocities are thus postulated to increase so that some do accelerate to speeds faster than c. However, Hoyle's theory does not pretend to explain the motion of the galaxies in terms of forces. It simply states a kinematic rule for calculating specified positions as a function of time, and the specifications are such that galaxies do disappear from view eventually. There is little point in searching for a force in nature which might result in Hoyle's expansion—It would have to be of an unfamiliar type, since no finite forces could ever result in such a motion in the framework of the present mechanics.

I have inherited a prejudice from my teacher, Dr. Wheeler, to consider it against the rules to explain a result by making a convenient change in the theory, when the present theory has not been fully investigated. I say this because there has been speculation that a force of expansion, which would tend to accelerate galaxies as Hoyle wants, would result if the positive charges were not exactly equal to the negative charges. Since we have an excellent theory of electrodynamics in which e^+ is precisely the opposite of e^-, such a speculation seems useless. We are not going to overthrow a beautiful theory of electrodynamics to provide a mechanism for a kinematic model which might easily be found wrong as soon as we make accurate observations. It is still quite possible for Hoyle to be right, but we shall find it out first of all from observations on the universe as it is.

Let me say also something that people who worry about mathematical proofs and inconsistencies seem not to know. There is no way of showing mathematically that a physical conclusion is wrong or inconsistent. All that can be shown is that the mathematical assumptions are wrong. If we find that certain mathematical assumptions lead to a logically inconsistent description of Nature, we change the assumptions, not Nature.

I say all this because I am not sure in which ways the theory of the Universe according to Hoyle may not quite coincide with many other assumptions we physicists ordinarily make. For example, there may be some trouble in the possibility of signalling which is so much a part of the thinking in relativistic theories. If a given galaxy has disappeared from view, is it really out of the universe, for us? Is it possible that we may ask a friend in the outer fringe how this galaxy is doing, while he is still in our view and the other galaxy is in his view? Or is there a general conspiracy of the detailed kinematics to preserve the relativistic rule, that a velocity near c in a coordinate system moving relative to us with a speed near c always yields a sum less than c?

Let us speak for a moment about the conservation of energy. In Hoyle's theory matter is created at rest at the "center" of the universe— and we have said that there is no over-all energy creation because the negative gravitational energy just balances the rest mass energy. This kind of conservation of energy is that if we take a box of finite size anywhere, no matter appears inside except by an energy flow from the outside. In other words, only a law of local conservation is meaningful. If energy could disappear in one place and simultaneously reappear in some other place, without the flow of anything in between, we could deduce no physical consequences from the over-all "conservation." Let us therefore interpret Hoyle's creation of matter as follows. We imagine a finite universe having large masses distributed in a spherical shell as illustrated in Figure 13.2. We imagine pairs of particles of zero energy dropping in from the outside of the shell toward the inside. These we might think of as photons or gravitons or neutrinos, or maybe some new particles, some shmootrinos which don't worry about baryon conservation. When they meet another shmootrino dropping in from the other side with opposite momentum, these can have enough energy to create a hydrogen atom. In this way we can have both Hoyle's matter creation and a local energy conservation, since the matter is created from energy flowing in by the shmootrinos. If the flux of shmootrinos were very high, and the cross section for matter production be finite just above threshold, we can understand why matter should be created at rest relative to the average of the nebulae. The idea is that if the flux is high enough, as soon as a shmootrino has enough energy it will find another coming in the opposite direction and proceed

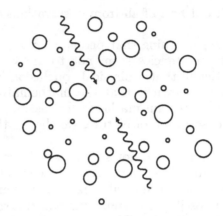

Figure 13.2

to create matter; if the maximum energy which shmootrinos can acquire in falling is barely the same as the threshold, the matter will be created "at rest." In some such manner we can have simultaneously local energy conservation and Hoyle's theory. Of course, there are a great many remaining problems, such as we still have to consider why baryon number may not be conserved.

13.4   MACH'S PRINCIPLE AND BOUNDARY CONDITIONS

The classical theory of gravitation has not led us to an answer to the question whether Mach's Principle is valid. We might ask, for example, whether the theory of gravitation predicts Coriolis forces if the whole of the galaxies have a net rotation about us. The problem has been approached in the following way. We imagine a rotating shell of matter at a large distance as in Figure 13.3 and ask whether there will be forces at the center to make a swinging pendulum follow the shell. The problem is solved putting in the boundary condition that $g_{\mu\nu} = \eta_{\mu\nu}$ at infinite distances. The result (obtained by Thirring [Thir 18]) is

$$\omega = \omega_0 \left(\frac{GM}{Rc^2} \right). \tag{13.4.1}$$

The quantity (GM/Rc^2) is always smaller than 1, so that the pendulum or whatever it is does not quite follow the rotating matter. It is always possible that this result may be prejudiced by the particular choice of boundary conditions. All that we know is that $g_{\mu\nu} \sim \eta_{\mu\nu}$ in regions where gravitational potentials are constant; there is no guarantee that $g_{\mu\nu} \sim \eta_{\mu\nu}$, if gravitational potentials are zero. Thus it may be that the apparent

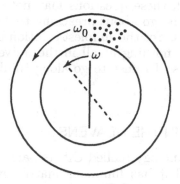

Figure 13.3

failure of Mach's Principle is due to our choice of a wrong boundary condition. We might compare the situation to a case in electrostatics; it is possible to solve the equations by setting $E_z = 1$ at large distances, as a boundary condition. We can extend these solutions to larger and larger regions, yet we are doing this at the expense of imagining large amounts of charge uniformly distributed along planes perpendicular to the z axis in the region outside. Setting $g_{\mu\nu} = \eta_{\mu\nu}$ at infinity may imply large amounts of matter uniformly distributed "outside of infinity." Thus it appears to me that we might learn whether Mach's Principle is consistent with our present theory by studying the meaning of the boundary conditions.

In our previous discussion of Mach's Principle we had speculated that perhaps the magnitude of $g_{\mu\nu}$ in the metric

$$(ds)^2 = g_{\mu\nu}\, dx^\mu\, dx^\nu \tag{13.4.2}$$

is a quantity of physical significance, if we measure the proper time in natural units such as Hubble's cT and lengths in some natural units such as Compton wavelengths; for proton wavelengths $g_{\mu\nu}$ is of magnitude 10^{-84}. Special relativity occurs as a special case $g_{\mu\nu} = $ constant, in which case the proper time and lengths can be scaled to laboratory measures. The nearby masses such as the sun contribute a small piece of $g_{\mu\nu}$, which is observed differently because locally it is rapidly varying compared to the variations of contributions from the nebulae. We have also seen how if we postulate an equal contribution from each baryon the order of magnitude of this 10^{-84} comes out. Is there something about this baryon number? For example, is it conserved? Is it infinite? Might not the number of baryons in an observable region of the universe have some consequence on physical processes?

The answer to all these questions may not be simple. I know there are some scientists who go about preaching that Nature always takes on the simplest solutions. Yet the simplest solution by far would be nothing, that there should be nothing at all in the universe. Nature is far more inventive than that, so I refuse to go along thinking it always has to be simple.

13.5 MYSTERIES IN THE HEAVENS

An old puzzle in cosmology (called Olbers paradox) has been, if the universe is infinite, and it has luminous matter everywhere, why are the heavens not infinitely bright? There is, of course, a real possibility that the number of stars is finite. If not, the heavens need not be infinitely bright—the absorption of light or the red shift and gravitational red shifts could all conspire to make even an infinite universe have a dark sky. Nevertheless the sky contains some objects of truly amazing intensity. The radio astronomers, whose instruments reach farther out than anyone else's, have found a number of radio sources having a peculiar structure. Figure 13.4 is a rough sketch of the contours of equal intensity. Some 20 or 25 of these things have been reported and charted. When the astronomers turn their telescopes to this region of space, they find in the center of this structure a galaxy, viewed from the edge. Occasionally one finds also radio sources whose contours of intensity have a single focus, and at the focus there appears a galaxy viewed broadside. This presumably represents an object of the same kind viewed from a different angle. What is truly amazing about these objects is the amounts of energy radiated away. When the radio astronomers integrate the intensity observed over *their* radio spectrum, making no extrapolations in the unobserved frequencies, they find an energy of some 10^{44} ergs/second is radiated away. If one estimates the lower limit for the lifetime of these objects by measuring its dimensions and dividing by the speed of light, one finds that the energy radiated away is of the order of 10^{60} ergs, or the equivalent of the rest masses of 10^6 to 10^8 stars of the size of our sun. The power radiated in visible light is an additional amount of similar size. This means that it were as if 10^6 to 10^8 stars had been completely annihilated! Where does such energy come from?

Ordinary nuclear processes which convert protons to iron can only annihilate a very small fraction of the mass. The number of stars in an ordinary galaxy is 10^9 on the average, so that 10^8 stars cannot be annihilated by nuclear processes in stars. We can make this comparison because the galaxies having these tremendously intense halos look just like other galaxies in visible light. Even exploding all the stars in an ordinary galaxy would hardly produce that much power. The only way to have such power radiated away from luminous objects would seem

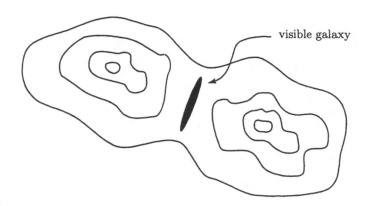

Figure 13.4

to be to have a million stars annihilate a million stars of antimatter. Alternative explanations involve some kind of structure at the center of these galaxies, some monster superstars in which the generation of energy follows paths very different from those of the ordinary star. The two-focus structure of Figure 13.4 is interpreted as being due to absorption of the radio frequencies by dark dust along the galactic plane.

One of the interesting facts about ordinary stars is that they do not vary greatly in size; their masses are always of the same order as that of our sun, possibly 10 times larger at most. It is not very difficult to establish that there may be a natural limit to the size of a star by considering simultaneously the requirements of the Pauli exclusion principle and escape velocities; we need to start filling in the electrons (say we do it at 0° Kelvin) in the Fermi sea and adding proton masses inside a given volume. At a mass of the order (1.5) the sun's mass, the top of the Fermi sea is critically high. Yet, the ordinary star is not sufficiently massive to make relativistic solutions very different from nonrelativistic solutions.

If we attempt to deal with the condensation of a superstar involving 10^9 sun masses, the gravitational processes might be highly relativistic. The pattern of cooling by radiation and collapse of the interior toward higher temperatures, which astrophysicists describe for ordinary stars, might not be valid. At sufficiently high pressures, the preferred direction of nuclear processes might be toward inverse β decays of the protons

$$p + e \rightarrow n + \nu. \qquad (13.5.1)$$

The evolution of a star of such high mass and rich in these neutrons might be quite different from our sun. I believe it is very necessary, before we make newer theories to explain these things, to make a serious attempt at using all the knowledge of our present physics to see what might happen under very peculiar circumstances.

14.1 THE PROBLEM OF SUPERSTARS IN GENERAL RELATIVITY

In this lecture I want to discuss a solution to the problem of superstars, those lumps of matter having masses of 10^9 suns which Fowler and Hoyle [HoFo 63] have been discussing recently. We take a model which is very simple, yet may possess a great many of the attributes of the real things. After we understand how to go about solving the simple problem, we may worry about refinements in the model. The starting point is the differential equation of general relativity, the Einstein equation.

$$8\pi \, G T_{\mu\nu} = G_{\mu\nu} = R_{\mu\nu} - \frac{1}{2} g_{\mu\nu} R \qquad (14.1.1)$$

The right-hand side is the "geometric" side; here we put in the curvatures in terms of the metric tensor. If we assume static, spherically symmetric solutions, then the elements of the metric tensor are specified by functions $\nu(r)$ and $\lambda(r)$ such that

$$(ds)^2 = e^{\nu}(dt)^2 - e^{\lambda}(dr)^2 - r^2\left(\sin^2\theta(d\phi)^2 + (d\theta)^2\right). \qquad (14.1.2)$$

The left-hand side of eq.(14.1.1) is the "physical" side, involving the stress-energy tensor. If we assume that the matter is gaseous, this tensor involves only the pressure p and the density ρ at any point. Denoting the coordinates (r, θ, ϕ, t) by indices in order $(1,2,3,4)$, and derivatives with respect to r by primes, the Einstein equation reduces to the following set, in terms of the functions $\nu(r)$ and $\lambda(r)$ and the pressure and density:

$$G^1{}_1 = -e^{-\lambda}\nu'/r - (e^{-\lambda} - 1)/r^2 = -8\pi Gp \tag{14.1.3a}$$

$$G^2{}_2 = -e^{-\lambda}\left(\nu''/2 - \nu'\lambda'/4 + (\nu')^2/4 + (\nu' - \lambda')/2r\right) = -8\pi Gp \tag{14.1.3b}$$

$$G^4{}_4 = -e^{-\lambda}\lambda'/r - (e^{-\lambda} - 1)/r^2 = 8\pi G\rho \tag{14.1.3c}$$

The model that we shall use will be specified by what we put in for p and ρ. These quantities represent the pressure and density which would be actually measured by an observer standing still at any particular point. We will not obtain valid solutions unless we take care that our physical tensor $T_{\mu\nu}$ satisfies the conservation laws. For our spherically symmetric case, only the radial component of the tensor divergence is of significance; we must have

$$\frac{\partial T^1{}_1}{\partial r} + \frac{1}{2}\nu'(T^1{}_1 - T^4{}_4) + \frac{1}{r}(T^1{}_1 - T^2{}_2) = 0 = -\frac{1}{2}\nu'(p + \rho) - p', \tag{14.1.4}$$

which states that the pressures in the radial direction are balanced, as should be in our static solution. The divergence equation serves to eliminate ν'. Next, we obtain a formula to eliminate $\exp(-\lambda)$. First we rewrite $G^4{}_4$ in terms of a new function $M(r)$, as follows

$$G^4{}_4 = \frac{1}{r^2}\frac{d}{dr}\left[r(1 - e^{-\lambda})\right]. \tag{14.1.5a}$$

If we set

$$M(r) = \frac{1}{2}\left[r(1 - e^{-\lambda})\right], \qquad e^{-\lambda} = 1 - \frac{2M(r)}{r}, \tag{14.1.5b}$$

then

$$\frac{dM}{dr} = 4\pi r^2 G\rho. \tag{14.1.5c}$$

The function $M(r)$ appears to be proportional to the mass of the star, since it is the integral of the density ρ. However its interpretation is not so direct, because of the peculiarities of the coordinates in which ρ is measured. We discuss this point later. Inserting expressions for ν' and $\exp(-\lambda)$ into eq.(14.1.3a) yields

$$\left(1 - \frac{2M}{r}\right)\frac{1}{r}\frac{dp}{dr} = -(p + \rho)\left(4\pi Gp + \frac{M}{r^3}\right). \tag{14.1.6}$$

Together with the differential equation for $M(r)$ and with an equation of state connecting p and ρ we have a set of coupled equations which may in principle be solved for $M(r)$, p and ρ; with suitable boundary conditions these may represent a superstar in the static approximation.

What kind of an equation of state shall we take? A mass of some 10^9 sun masses is still very dilute when spread over a region of galactic dimensions: even at a temperature of some 10^9 degrees Kelvin, the gas pressure is rather low. However, the radiation density, which is proportional to T^4, turns out at such temperatures to be a significant fraction of the rest-mass energy of the nucleon density. We obtain a meaningful approximation by neglecting the gas pressure in comparison with the radiation pressure: in the same spirit, we neglect the small increase in mass of the nucleons because of their velocities. In units of the nucleon rest mass energy, we have then, if s is the density of nucleons,

$$\rho = s + \epsilon \qquad (14.1.7a)$$

$$p = \frac{1}{3}\epsilon \qquad (14.1.7b)$$

These equations connect p and ρ but we still need to specify ϵ in order to have an equation of state. We make the adiabatic assumption —which everyone makes in trying to deal with such problems—that the temperature distribution is the same as though the whole thing had fallen together from an initially uniform distribution, without any mixing or energy transfer between different regions. As we compress the matter inside of a box, all frequencies inside go up by the same factor inversely as the length of the box. Since the entropy is constant for an adiabatic process, the temperature must rise in the same way. The density of nucleons is thus proportional to the cube of the temperature, and the radiation energy density is proportional to T^4. In terms of the temperatures measured in units of 10^9 degrees, and energies in units of the nucleon rest mass energy,

$$\epsilon = aT^4, \qquad s = a\tau T^3. \qquad (14.1.8)$$

The number am_n, where m_n is the nucleon mass, is a constant having the value 8.4 g/cm^3; τ is a parameter related to the star's radius-independent entropy per baryon by (entropy per baryon) $= 4/(3\tau)$. These results may be deduced also from the general condition for adiabatic compression, which may be expressed as

$$s^2 \frac{d(\epsilon/s)}{ds} = p = \frac{\epsilon}{3}. \qquad (14.1.9)$$

These interrelations between pressures and densities and adiabatic processes have been worked out in connection with stellar problems in the classical case. Stars in which both pressure and density follow power laws in the temperature everywhere are known as polytropes.

In terms of a new temperature, $t = T/\tau$, and new units such that $8\pi G m_n a \tau^4 = 1$, the set of equations reduces to the following.

$$\rho = a\tau^4 \left[t^4 + t^3 \right] \tag{14.1.10a}$$

$$p = a\tau^4 \left[\frac{1}{3} t^4 \right] \tag{14.1.10b}$$

$$\frac{dm}{dr} = \frac{1}{2} \left[t^4 + t^3 \right] r^2 \tag{14.1.10c}$$

$$\frac{dt}{dr} = -\frac{r}{2}\left(\frac{3}{4} + t \right)\left(\frac{1}{3} t^4 + \frac{2m}{r^3} \right) \Big/ \left(1 - \frac{2m}{r} \right) \tag{14.1.10d}$$

What shall we choose as the boundary conditions? We assume a certain central temperature and that the surface is very much cooler, essentially zero in comparison. The input numbers to find the solutions $t(r)$ and $m(r)$ are simply the following:

$$m = 0, \qquad t = t_c, \qquad \text{at} \quad r = 0. \tag{14.1.11}$$

The problem is now stated in such a fashion that numerical solution is very simple. We start at the center, where we know $m(r) = 0$, and $t(r) = t_c$; we compute (dm/dr) by eq.(14.1.10c), and compute (dt/dr) with eq.(14.1.10d), and then jump back and forth between these equations to generate $m(r)$ and $t(r)$. Since the derivative (dt/dr) will always be negative for t positive, at some point r_0, t becomes zero. We stop the solution at this point, and assume that a more physical solution would change only the outermost layers of the star, to make it taper off smoothly toward zero density without changing the solution in the interior to any great extent. Thus, the radius r_0 is assumed to represent the radius of the star, and the value $m_0 = m(r_0)$ to represent the total mass of the star.

14.2   THE SIGNIFICANCE OF SOLUTIONS AND THEIR PARAMETERS

The solution we have described is valid for many kinds of stars, that is, stars described by all values of the parameter τ. In order to give an idea of the magnitudes of m_0 and r_0 for cases of interest, we give the conversion factors to more familiar units:

$$M \equiv \text{Mass of the star } = (27 \times 10^6 \text{ sun masses}) 2m_0/\tau^2 \quad (14.2.1a)$$
$$R \equiv \text{Radius of the star } = (8 \times 10^{12} \text{cm}) r_0/\tau^2 \quad (14.2.1b)$$
$$T_c \equiv \text{Central Temperature } = t_c \tau (10^9 \text{degrees}) \quad (14.2.1c)$$
$$M_{\text{rest}} \equiv \text{Rest mass of the star's nucleons}$$
$$= (27 \times 10^6 \text{ sun masses}) 2N/\tau^2 \quad (14.2.1d)$$

There are various ways in which we can see that our equations describe something that our intuition approves. For example, in the case that the mass $m(r)$ never gets very large, the pressure varies with the radius according to the Newtonian rule,

$$\frac{dp}{dr} = -\frac{m(r)}{r^2} \rho. \quad (14.2.2)$$

An interesting point concerns the total number of nucleons. Although we may be tempted to write simply $4\pi \int dr \, sr^2$, we ought to remember to write down properly invariant expressions. The true value is

$$N = \int_0^{r_0} s^4 \sqrt{-g} \, dr \, d\theta \, d\phi, \qquad \sqrt{-g} = e^{\nu/2} e^{\lambda/2} r^2 \sin\theta, \quad (14.2.3)$$

where s^4 is the time component of a four vector, s^μ. We may compute its value and carry out the integration in a system in which the nucleons are at rest; in this system, only the time component is nonzero, so that we conclude

$$s^\mu s_\mu = (s)^2 = g_{44} s^4 s^4, \qquad s^4 = s \, e^{-\nu/2}. \quad (14.2.4)$$

The result for the total number of nucleons is therefore

$$N = 4\pi \int_0^{r_0} dr \, sr^2 [1 - 2m/r]^{-1/2}. \quad (14.2.5)$$

Let us now look again at the expression for the mass of the star, and try to understand it more fully. The density ρ is a sum of two terms, a rest mass energy s and a radiation energy ϵ. When we write the mass explicitly as an integral over properly invariant volume elements, we see that the density ρ is multiplied by a square root factor.

$$m \leq m_0 = 4\pi \int_0^{r_0} \frac{r^2 dr}{\sqrt{1 - 2m/r}} \rho \sqrt{1 - 2m/r} \quad (14.2.6)$$

This is precisely what we should have expected from a relativistic theory —the square root corrects the energy density by the gravitational energy.

Table 14. 1

t_c	r_0	$2m_0$	$2N$
0.01	1100	8.45	8.45
0.10	114	6.23	6.19
0.20	59	4.71	4.62
0.40	36	3.13	2.97
0.60	28	2.39	2.19
1.00	32	1.87	1.66

When the temperatures get to be higher than 10^9 degrees, we must beware in attempting to use these solutions, lest new physical processes which can occur at these high temperatures make our equation of state completely inadequate. For example, if neutrino pairs can be produced from electron-electron collisions, they may carry away large amounts of energy so that our approximations may be totally invalid. These processes might begin to be important at 10^9 degrees, a temperature hot enough that a significant fraction of the particles have sufficient kinetic energy to make electron pairs. The possibility of making these pairs will change the relations between s and ϵ, and between ϵ and p. However, in the adiabatic approximation, the connections are completely specified by a quantity γ, (the pressure p is proportional to s^γ), and it is found that γ does not vary very much as the temperature is changed. It has the same value at both extreme limits; $T \to 0$ and $T \to \infty$ both have $\gamma = 1.333$. There is a minimum in between, at which $\gamma = 1.270$. This suggests that the corrections due to the electron pairs will not alter the qualitative aspects of our answers.

14.3 SOME NUMERICAL RESULTS

Preliminary calculations have yielded the results listed in Table 14.1, for a given value of τ, varying only the central temperature. The table lists the temperature t_c, the radius r_0, the mass of the star m_0 and the "number of nucleons N." What is interesting about these numbers? In order to find out what happens to a given star, we might ask what is the sequence of radii and central temperatures corresponding to the same number of nucleons, if the total energy per nucleon is going down. We expect this to simulate the situation in a star which is slowly radiating energy away. The numbers in the table are not comprehensive enough that we may be sure, yet we see that the energy per nucleon goes down as the central temperature goes down; that is, the star actually cools off as it radiates energy away.

Will these stars spontaneously spread out? The stability of our star has not been studied. Within the framework of the same solution, computations which result in the same number of nucleons and the same τ may be compared as to radii and central temperature. The fact that there is apparently a minimum in the radius for t_c somewhere between 0.40 and 1.00 is very suggestive; the star may have a stable position. Another way of studying the stability is to consider "explosions." Suppose that we compute the total energy of a number N of nucleons with a certain entropy per nucleon, i.e., a certain τ, and then break up the nucleons into two stars of the same τ, keeping the sum N constant. Do we get work out of the process, or did we have to do work to get it broken up? The τ is supposed to be the same, because all of the material is assumed to fall together from the same initial distribution. Can we find anything about this from the numbers? As N decreases, we find that the excess mass is *rising*. This means that the two objects of smaller N would be more massive—work is required to break up the system. This suggests that a star might not throw off material, but keep together in one lump.

The preceding considerations also show us that stars which follow our model could not actually be formed; they all have more energy than the rest mass of the nucleons, therefore it took some energy to put them together.

One of the things we may have found is that in any case, the corrections due to general relativity are considerable, and very important. In what direction do the electron pairs push the solution? They tend to make a star even more like those which get hotter at the center as they cool off from the outside.

Do we ever get into trouble with the gravitational radius? We have set up our equations so that $m(r) = 0$ at the origin, and it increases outward. If we ever got to large masses such that $2m(r) = r$ very nearly, our differential equation (14.1.10d) shows that near the critical value of r, t would tend to $-\infty$ logarithmically. Thus, before that point the temperature would drop to zero, and in our scheme we would stop our solution at this point. Nevertheless, the numerical results for the mass and radius are so far from being critical that perhaps we need not worry about this problem at present.

14.4   PROJECTS AND CONJECTURES FOR FUTURE INVESTIGATIONS OF SUPERSTARS

There is another mathematical definition of the star problem which may be amenable to treatment. We have found that the total number of nucleons in the star is given by

$$N = 4\pi \int_0^{r_0} \frac{sr^2 \, dr}{\sqrt{1 - 2m/r}}, \qquad (14.4.1)$$

where

$$m(r) = 4\pi \int_0^r dr' \, \rho \, r'^2,$$

and $\rho = \rho(s)$ is known in terms of a postulated equation of state such as our "adiabatic"

$$s \frac{d\rho}{ds} = p + \rho. \qquad (14.4.2)$$

The equilibrium problem tries to determine the configuration of least mass starting with a fixed number of nucleons. We can get the same information by fixing the mass and asking for the maximum number of nucleons. The mathematical formulation is as a variational differential functional equation

$$\frac{\delta N}{\delta s(r)} = 0. \qquad (14.4.3)$$

If we manage to solve it, we obtain extremum solutions $s(r)$. It is comforting to me to feel that even very complicated problems are trying to look simple, in terms of appropriate principles! We will have found the minimum-mass solutions if the extremum $N[s(r)]$ is truly a maximum.

After we have investigated static solutions, we may turn our attention to the full dynamical problem. The differential equations are horrifying. As one looks at its enormously complex structure and begins to make comparisons to the classical limit, the significance of many of the terms becomes more apparent. In the simplest case of gas dynamics, the equations describe the propagation of sound in inhomogeneous media; it is nonlinear sound, so that it can form shock waves, etc. No wonder the thing is so complicated. A more modest investigation might concern small vibrations about the static solutions; real frequencies would denote that our previous solutions were truly stable, once formed, and imaginary frequencies would tell us that our solutions were unstable.

A refined calculation needs also better expressions for the "physical" side of the equations. What happens if we take neutrinos out of the center of the star? Will it continue to fall in, or what? In case that the star is highly relativistic, then these neutrinos may remove a large fraction of the total energy, and thus make a significant reduction in the gravitational attraction. The theory of stars is on fairly firm ground classically, when the rest mass of the particles accounts for nearly all of the energy. In this case, removing energy from the center of the star results in a further collapse, which makes the center hotter. As the center gets hotter, the nuclear reactions furnish more energy, which must be carried away if the star is to remain stable. If the center has gotten hot enough that the lighting of the nuclear fuel produces more energy than can be carried away, the situation becomes unstable, and the star explodes. In the highly relativistic case, however, new qualitative features begin to appear when the radiation energy becomes a large fraction of the total mass. Here, as the center "cools" by loss of energy, it also becomes less attractive, because a sizeable fraction of the mass is removed. Thus, it may be that for a sufficiently large mass there may be no process leading to an explosion.

I expect that the solutions to the problem will show that for masses larger than some 10^8 sun masses, spherically symmetric solutions for condensing matter do not lead to a collapse, but sort of slosh in and out about a certain most favorable radius. The usual processes of stellar development can take place if the distribution becomes unspherical. In this direction, then, perhaps we can find an explanation to the fact that all visible stars appear to be very nearly of the same size. A solution of the full dynamical problem may lead us to understand how matter uniformly distributed may begin to condense symmetrically, and then at a certain point prefer to form bumps which may condense further. The results may turn out to be highly sensitive to any amount of angular momentum originally possessed by the condensing mass. For example, the planets contain nearly 95% of the total angular momentum in our solar system. It may be that a condensing mass can form globs to which it transfers most of the angular momentum.

15.1 THE PHYSICAL TOPOLOGY OF THE SCHWARZSCHILD SOLUTIONS

In the preceding lecture we have had some hints that distributions of real matter cannot condense to a radius smaller than the gravitational radius $2m$; even if we tentatively conclude that "wormholes" cannot be formed from real matter, the question remains as to whether the Schwarzschild solution truly represents a case in which the tensor $G^{\mu}{}_{\nu}$ is zero *everywhere*, a case in which no matter at all can look like matter when viewed from a distance. Let us therefore attempt to continue the solutions of Schwarzschild inside the critical radius $2m$. We suspect that this must be possible because although the metric

$$(ds)^2 = (1-2m/r)(dt)^2 - \frac{(dr)^2}{(1 - 2m/r)} - r^2\left((d\theta)^2 + \sin^2\theta(d\phi)^2\right) \quad (15.1.1)$$

has an apparent singularity at $r = 2m$, the curvatures are smooth at this point. The curvatures are singular at the origin, $r = 0$, so that something terrible is truly happening to the space at the origin. A spaceship falling into the origin might be catastrophically distorted because the tidal forces become infinite—this is the kind of horrible behavior that singular curvatures imply. All that happens at $r = 2m$ is that the coefficients of $(dt)^2$ and of $(dr)^2$ change sign in eq.(15.1.1)—yet the space is still three-and-one, so it might feel quite normal.

Let us consider an expansion of the space about the singular point. Suppose that we change coordinates in the neighborhood of $r = 2m$, and consider the planes $d\phi = 0, d\theta = 0$. In terms of a new variable x, we have

$$x = (1 - 2m/r),$$
$$r = 2m(1 + x) \qquad \text{for } x \text{ small,} \tag{15.1.2}$$
$$(ds)^2 = x(dt)^2 - (2m)^2 \frac{(dx)^2}{x},$$

near the singular point. Although this space reverses sign as x changes sign, for $x > 0$ the metric can be changed again so that it becomes flat; a simple coordinate transformation reduces it to a polar form

$$x = R^2 \rightarrow (ds)^2 = R^2(dt)^2 - (4m)^2(dR)^2, \tag{15.1.3}$$

which easily transforms into the Minkowski metric by means of the substitution

$$v = 4mR \cosh(t/4m),$$
$$u = 4mR \sinh(t/4m), \tag{15.1.4}$$
$$\rightarrow (ds)^2 = (du)^2 - (dv)^2.$$

These results demonstrate that the space near the singular point is perfectly well behaved, so that the Schwarzschild singularity is a peculiarity of the coordinates that we have defined. In order to connect the geodesics across the point $r = 2m$, the equation (15.1.4) suggests the substitution,

$$x = \left(1 - \frac{2m}{r}\right) = -\frac{(u^2 - v^2)}{(4m)^2}, \qquad \frac{u}{v} = \tanh\left(\frac{t}{4m}\right), \tag{15.1.5}$$

in terms of the coordinates u and v, the space and the metric are smooth on both sides of $r = 2m$. A similar substitution has been used by Fuller and Wheeler [FuWh 62] to get across the gap. The geodesics properly joined across $r = 2m$ show that particles falling into a mass smaller than its critical radius $2m$ do not get reflected into any "new" space on the other side of any throat, but keep right on falling toward the origin. There is no contradiction here with the considerations which led to the speculations on the wormholes. The throat-like topology was obtained by slicing up the space in a particular way, setting $dt = 0$. However, the motion of real particles does not occur in a space in which $dt = 0$, and there is no reason why the topology of the subspace $dt = 0$ should correspond to a general property of the four-dimensional space. A toroidal doughnut can be cut from a solid piece—yet there is nothing toroidal about the solid piece. For physical problems, the topology of interest concerns the geodesics—and there are no time-like geodesics which go through the wormhole.

15.2   PARTICLE ORBITS IN A SCHWARZSCHILD FIELD

It is instructive to solve for the radial motion of particles as a function of the proper time s. As usual in central-force problems, the motion happens in a single plane (we take it to be $\theta = \pi/2$), and the radial motion is determined by two parameters, K and L, related to the total energy and the angular momentum, which are first integrals of the time equation and of the angle equation, as follows: The geodesic equations,

$$\frac{d}{ds}\left(g_{\mu\nu}\frac{dx^\mu}{ds}\right) = \frac{1}{2}\frac{\partial g_{\alpha\beta}}{\partial x^\nu}\frac{dx^\alpha}{ds}\frac{dx^\beta}{ds}, \tag{15.2.1}$$

may be trivially integrated when $\nu = 3, 4$ (coordinates ϕ, t) because the metric tensor is independent of ϕ and t, and therefore the right-hand side of eq.(15.2.1) is zero. This defines the following integrals:

$$K = (1 - 2m/r)\frac{dt}{ds}, \qquad L = r^2\frac{d\phi}{ds}. \tag{15.2.2}$$

The radial equation may be obtained from setting $\nu = 1$ in eq.(15.2.1), but this is more work than is necessary. It is easier to obtain the radial equation from the condition

$$g_{\mu\nu}\frac{dx^\mu}{ds}\frac{dx^\nu}{ds} = 1, \tag{15.2.3}$$

which may be explicitly written, in terms of the quantities L and K as follows:

$$\frac{K^2}{(1 - 2m/r)} - \frac{1}{(1 - 2m/r)}\left(\frac{dr}{ds}\right)^2 - \frac{L^2}{r^2} = 1. \tag{15.2.4}$$

The proper time elapsed as a particle falls from one radius r_0 to another r_1, is given by

$$\int ds = \int_{r_0}^{r_1} dr\left(K^2 - (1 - 2m/r)(1 + L^2/r^2)\right)^{-1/2}. \tag{15.2.5}$$

One thing to notice is that nothing terrible happens at $r = 2m$ anymore; the integrand is well behaved and there is no joining problem across any gap. If we had studied orbital motions in the first place, and not worried about the metric, we might not have noticed the Schwarzschild singularity, and we would have got the right answers by simply using (15.2.5).

The significance of the square root is the usual one of orbit motions. The integral is stopped if the square root becomes negative—smaller values of the radius are never reached by the particle. If the angular momentum L is sufficiently large, the square root becomes imaginary at a radius larger than $2m$, and the orbits have the same qualitative behavior as in the Newtonian case. On the other hand, if the energy and angular momentum are such that the particle does cross the radius $2m$, the square root does not become negative inside, which means that all particles continue to fall toward the origin. In fact, once inside $r = 2m$, particles of larger angular momentum L fall faster, the "centrifugal force" apparently acts as an attraction rather than a repulsion.

At this point I want to mention some peculiar results which are obtained when one assumes that the Schwarzschild field corresponds to a charged object when viewed from a distance. It may readily be shown that the only change in the metric consists in the following replacement

$$(1 - 2m/r) \rightarrow (1 - 2m/r + q^2/r^2) \qquad (15.2.6)$$

where q is the apparent charge. When such an expression is inserted into the proper time interval eq.(15.2.5), the square root is inevitably imaginary for sufficiently small radius, so that particles never would fall into the origin but would always be reflected back out. This repulsion is not due to an electrical force between the particles—it is inherent in the metric if we insist that the fields should correspond to those of a charged particle at the origin, for large radius r. Thus, the repulsion would be felt even by a neutral particle falling into a charged center.

The metric corresponding to a charged mass eq.(15.2.6) evidently has *two* singular points. It would be of some interest to study the continuation of the geodesics of a falling particle through these two singularities; it is not inconceivable that it might turn out that the reflected particle comes out earlier than it went in! I suspect this because apparently the falling particle would take infinite time to reach the first singularity (from the viewpoint of someone outside), yet the whole trajectory in and out from the point of view of the particle itself would take a finite time.

15.3   ON THE FUTURE OF GEOMETRODYNAMICS

The many discussions we have had on the Schwarzschild solutions are a symptom of the fact that we have a theory which is not fully investigated. It is time we passed on to other subjects, yet I want to give you my guesses as to what the answers will be, once the theory is studied. The original speculations of J. A. Wheeler on wormholes were based on the idea that it might be possible to construct solutions of Einstein's equations for which $G^\mu{}_\nu = 0$ everywhere, which would nevertheless act and feel as though they were real masses. The topology of wormholes is such that it was intuitively clear that electric field lines going into a wormhole and coming out of a wormhole somewhere else would very nicely correspond to the existence of positive and negative charges of exactly the same size. Even though we have demonstrated that the topology of the geodesic space is not that of a wormhole, the idea that matter and charge may be manifestations of the topology of space is very beautiful and exciting, and it is by no means discredited simply because it does not yield any quantitative result in terms of the Schwarzschild solution. It would indeed be very beautiful to have $G^\mu_\nu = 0$ everywhere, so that, in words used recently to describe geometrodynamics, matter comes from no matter, and charge comes from no charge.

In the immediate future, one should investigate the properties of the Schwarzschild solution at the origin, $r = 0$. I suspect that it will not be possible to demonstrate that $G^\mu_\nu = 0$ *everywhere*, but rather that $G^\mu_\nu = \delta(x)$, or something of the kind. The explanations of the behavior of charges will require further detailed study; I believe that the "repulsion" at the origin will turn out to be an erroneous conclusion, due in general to the inconsistency in assuming a point charge; the energy density about a point charge goes as E^2 or $1/r^4$, which implies that the mass inside any finite radius should be infinite. If the mass inside is not to be infinite, we must write something like

$$\text{Mass Inside} = (\text{Constant}) - \frac{q^2}{2r}. \tag{15.3.1}$$

If there is to be no negative mass inside any radius, then we are not allowed to go inside a region of radius a where a is defined by requiring $(\text{Constant}) = q^2/2a$. The magnitude of this constant would be arbitrary if the mass is not to be purely electromagnetic in origin. In the region outside a, we would have the following gravitational field and potential

$$\text{Field} = -\frac{q^2}{2r^2}\left(\frac{1}{a} - \frac{1}{r}\right), \qquad \text{Potential} = \frac{q^2}{2r}\left(\frac{1}{a} - \frac{1}{2r}\right). \tag{15.3.2}$$

By making the constant non-infinite and explicit, we cannot have a repulsion. In terms of new coordinates u and v the entire real world is contained in a subregion, and the geodesics of falling particles run into a *barrier* at $r = 0$. The originally worrisome gap at $r = m + \sqrt{m^2 - q^2}$ corresponds to a perfectly well behaved region of space, where the geodesics don't even have a kink. (See Figure 15.1.) It would be of great interest to study the geometry of this charge mass as we make it smaller and smaller.

Although geometrodynamics as developed by J. A. Wheeler and his co-workers has not yielded any quantitative results, it contains seeds of bold imagination which may yet lead to spectacular successes in our understanding of physics. One must give credit to Wheeler for really knowing

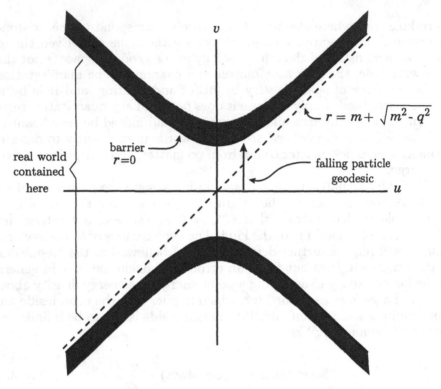

Figure 15.1

those features of our present theories which are unexplored and promising. During the time I was his assistant I benefited more than once from
his intuitive genius for knowing in which direction the answer lay. At one
time I was trying to construct a theory of classical electrodynamics in
which charges interacted only with other charges, instead of interacting
with fields—I felt the fields should disappear except as a way of keeping
track of the retardation. All went along very well until the time came to
explain the radiation reaction—in which a force is felt by an accelerating
particle long before its fields have had time to travel to other charges and
back. When I told Wheeler of my problems, he said "Why don't you use
the advanced potential?" The Advanced Potential? That was something
which everyone threw away as useless. It was obviously devoid of physical meaning—to suggest its use was magnificently daring. Yet some time
later the quantitative theory of its use was worked out—and we had a
theory of electrodynamics in which charges acted only on other charges,
by using for the potential, one half of the retarded, and one half of the
advanced.

On another occasion I received a phone call from him in the middle
of the night, when he said to me "I know why all electrons and positrons
have the same charge!" Then he explained further, "They are all the same

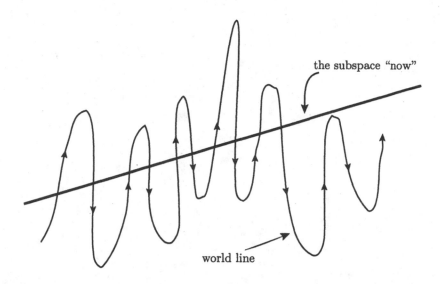

the subspace "now"

world line

Figure 15.2

electron!" His idea was that if the same object has a world line which is extremely complicated, when we look at it in the subspace "now," we see it at many different places. (See Figure 15.2.) Later on, I was able to make this kind of idea quantitative, by interpreting a positron as being an electron whose phase is going backward in time—and developing simplified methods for calculating matrix elements involving annihilation and creation of pairs. It would truly be very beautiful if the idea of wormholes, and geometrodynamics, could be perfected to improve our understanding of Nature—and knowing Wheeler, it does not seem to me unlikely that his intuition may be vindicated some day.

With these comments on problems of current interest we end the discussion of the classical theory of gravitation.

16.1 THE COUPLING BETWEEN MATTER FIELDS AND GRAVITY

In Lecture 10, we have written down the action terms corresponding to the propagation of free particles and fields. All that is left over from the complete action may be considered as a coupling between the fields, and we may proceed to compute various processes by perturbation theory. There needs to be no apology for the use of perturbations, since gravity is far weaker than other fields for which perturbation theory seems to make extremely accurate predictions. The known parts of the general action are the following:

$$-\frac{1}{2\lambda^2}\int d^4x \sqrt{-g}\, R + \frac{1}{2}\int d^4x \sqrt{-g}\left(g^{\mu\nu}\phi_{,\mu}\phi_{,\nu} - m^2\phi^2\right)$$
$$- \alpha \int d^4x \sqrt{-g}\, R\phi^2. \quad (16.1.1)$$

The first simplification we shall make is to set the coefficient α equal to zero. Leaving such a term in the action ordinarily worsens many problems of divergence that we shall encounter later, and it increases the labor of computing. Since any choice of the coefficient would be arbitrary at the present state of the experimental art, we choose the value which simplified the arithmetic most conveniently. The second step is to pull out the term representing the propagator of the fields by introducing the expansion

$$g_{\mu\nu} = \eta_{\mu\nu} + 2\lambda h_{\mu\nu}. \tag{16.1.2}$$

When we write down the action in terms of the fields $h_{\mu\nu}$ and scalar matter fields ϕ, we obtain the following:

$$\text{Action} = \int d^4x \; F^2[h_{\mu\nu}] + \int d^4x \; I[h_{\mu\nu}, \phi] + \int d^4x \; M[\phi], \tag{16.1.3}$$

where

$$F^2[h_{\mu\nu}] = \frac{1}{2}\left[h^{\mu\nu,\lambda}\overline{h}^{\mu\nu}{}_{,\lambda} - 2\overline{h}^{\mu\lambda}{}_{,\lambda}\overline{h}_{\mu\nu}{}^{,\nu} \right]$$

$$M = \frac{1}{2}\left(\eta^{\mu\nu}\phi_{,\mu}\phi_{,\nu} - m^2\phi \right).$$

The variations of the function I with respect to the fields $h_{\mu\nu}$ or ϕ represent the source terms in the differential equations of the fields. These may be written as follows, in space and in momentum representations:

$$\Box^2\phi - m^2\phi = -\left(\frac{\delta I}{\delta\phi}\right) \;\rightarrow\; \phi = -\frac{1}{(k^2 - m^2 + i\epsilon)} I\left(\frac{\delta I}{\delta\phi}\right)$$

$$-h_{\mu\nu,\lambda}{}^{,\lambda} + \overline{h}_{\mu\lambda,\nu}{}^{,\lambda} + \overline{h}_{\nu\lambda,\mu}{}^{,\lambda} = \lambda S_{\mu\nu} \quad \text{where} \quad S^{\mu\nu} = -\frac{1}{\lambda}\left(\frac{\delta I}{\delta h_{\mu\nu}}\right). \tag{16.1.4}$$

Note that $S_{\mu\nu}$ is what we called $^{\text{new}}T_{\mu\nu}$ in Lecture 6 (see Eq. (6.1.2)). How do we go on from here? Because of the careful design of the original action, as an invariant integral, it any be shown that the ordinary divergence of the source tensor $S_{\mu\nu}$ is identically zero. In the momentum representation,

$$k^\nu S_{\mu\nu} = 0. \tag{16.1.5}$$

The source tensor contains appropriately both the matter sources and the gravity sources. Because of the freedom we have in choosing a gauge, we can make the barred tensor $\overline{h}_{\mu\nu}$ divergenceless, and thus obtain a solution,

$$k^\nu\overline{h}_{\mu\nu} = 0 \rightarrow k^2 h_{\mu\nu} = \lambda \overline{S}_{\mu\nu}, \qquad h_{\mu\nu} = \frac{\lambda}{k^2 + i\epsilon}\overline{S}_{\mu\nu}. \tag{16.1.6}$$

The tensor on the right is not simply an unknown source tensor, but it is now well defined in terms of the original action eq.(16.1.1) and the expansion eq.(16.1.2), so that the equations are properly consistent and energy is conserved. Once we have an expansion in powers of the coupling constant λ, then we can proceed by the familiar rules of perturbation theory to compute all the diagrams to any given order in λ. The key expansions are those of $g^{\mu\nu}$ and of $\sqrt{-g}$. The first is easily written down by analogy with the expansion of $(1+x)^{-1}$ when x is a small number. We have

$$g^{\mu\nu} = \left(\eta_{\mu\nu} + 2\lambda h_{\mu\nu}\right)^{-1} = \eta^{\mu\nu} - 2\lambda h^{\mu\nu} + 4\lambda^2 h^\mu{}_\beta h^{\beta\nu} - 8\lambda^3 h^{\mu\beta} h_{\beta\tau} h^{\tau\nu} + \dots,$$
(16.1.7)

where one must remember the flat-space summation convention, as in eq.(4.1.6). The formula for $\sqrt{-g}$ may be calculated by means of the tricks of Lecture 6. Using eq.(6.3.11) with

$$g_{\mu\nu} = \eta_{\mu\beta}\left(\delta^\beta{}_\nu + 2\lambda h^\beta{}_\nu\right),$$

we have

$$\sqrt{-\mathrm{Det}\, g_{\mu\nu}}$$

$$= \sqrt{-\mathrm{Det}\, \eta_{\mu\beta}}\, \exp\left[\frac{1}{2} \mathrm{Tr} \log(\delta^\beta{}_\nu + 2\lambda h^\beta{}_\nu)\right]$$

$$= \exp\left[\frac{1}{2}\mathrm{Tr}\left(2\lambda h^\beta{}_\nu - \frac{1}{2}(2\lambda)^2 h^\beta{}_\tau h^\tau{}_\nu + \frac{1}{3}(2\lambda)^3 h^\beta{}_\tau h^\tau{}_\sigma h^\sigma{}_\nu + \dots\right)\right]$$

$$= \exp\left[\frac{1}{2}\left(2\lambda h^\beta{}_\beta - \frac{1}{2}(2\lambda)^2 h^\beta{}_\tau h^\tau{}_\beta + \frac{1}{3}(2\lambda)^3 h^\beta{}_\tau h^\tau{}_\sigma h^\sigma{}_\beta + \dots\right)\right]$$

$$= 1 + \lambda h^\beta{}_\beta - \lambda^2\left(h^\beta{}_\rho \overline{h}^\rho{}_\beta\right) + \dots.$$
(16.1.8)

Plugging these expressions for $\sqrt{-g}$ and for $g^{\mu\nu}$ into the action, we get explicit forms for the coupling of matter and gravity; the result from the second term of eq.(16.1.1) is, for example

$$S_m = \frac{1}{2} \int \left[\left(\eta^{\mu\nu} - 2\lambda h^{\mu\nu} + (2\lambda)^2 h^{\mu\beta} h_\beta{}^\nu + \dots\right)(\phi_{,\mu}\phi_{,\nu}) - m^2\phi^2\right]$$

$$\left(1 + \lambda h^\rho{}_\rho - \lambda^2(h^\sigma{}_\rho \overline{h}^\rho{}_\sigma) + \dots\right) d^4x$$

$$= \frac{1}{2}\int d^4x(\phi^{,\mu}\phi_{,\mu} - m^2\phi^2) - \lambda \int d^4x\, h^{\mu\nu}\left[\overline{\phi_{,\mu}\phi_{,\nu}} + \frac{1}{2}m^2\phi^2\eta_{\mu\nu}\right]$$

$$- \lambda^2 \int d^4x\left[\frac{1}{2}h^\lambda{}_\rho \overline{h}^\rho{}_\lambda(\phi^{,\mu}\phi_{,\mu} - m^2\phi^2) - 2h^{\mu\rho}\overline{h}_\rho{}^\nu\phi_{,\mu}\phi_{,\nu}\right].$$
(16.1.9)

The lowest order terms involve the interaction of two ϕ's and one h; this corresponds to a vertex such as is shown in Figure 16.1(a). At each vertex,

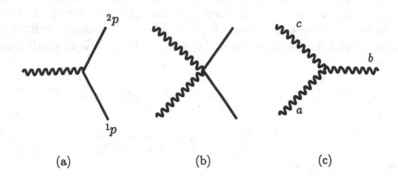

(a)   (b)   (c)

Figure 16.1

we require that the linear momenta be conserved. This rule comes from the volume integration in the action—there is no contribution unless the total phase of the term is equal to zero. We write plane wave solution as:

$$h_{\mu\nu} = e_{\mu\nu}\, e^{iq\cdot x}, \qquad \phi = e^{ip\cdot x}; \qquad (16.1.10)$$

in terms of the polarization tensors $e_{\mu\nu}$, the amplitude at a first order vertex is

$$-2\lambda \left[e^{\mu\nu}\, {}^1p_\mu\, {}^2p_\nu - \frac{1}{2} e^\rho{}_\rho \left({}^1p_\tau\, {}^2p^\tau - m^2 \right) \right]. \qquad (16.1.11)$$

Any diagram which involves only such vertices may now be calculated by simply plugging in the proper amplitudes at each vertex and the particle and graviton propagators between vertices, just as in electrodynamics.

Let us now look at the next order. The terms shown in eq.(16.1.9) involve products of two h's and two ϕ's, so that two lines and two wiggles come together at a junction, as in Figure 16.1(b). There are also terms coming from the expansion of the first term in eq.(16.1.1), involving products of three h's, corresponding to diagrams in which three wiggles come together at a point, as in Figure 16.1(c). The profusion of implicit sums in three indices results in terms which are very, very long when written out explicitly. For example, one of the terms in which three wiggles come together is $h_{\mu\nu,\beta}h^{\mu\beta}h^{\nu\alpha}{}_{,\alpha}$; when this is translated to momentum and polarization components, we get terms corresponding to all permutations of the three gravitons, for example

$${}^aq_\beta\, {}^ae_{\mu\nu}\, {}^be^{\mu\beta}\, {}^cq_\alpha\, {}^ce^{\nu\alpha} + {}^bq_\beta\, {}^be_{\mu\nu}\, {}^ae^{\mu\beta}\, {}^cq_\alpha\, {}^ce^{\nu\alpha} + {}^bq_\beta\, {}^be_{\mu\nu}\, {}^ce^{\mu\beta}\, {}^aq_\alpha\, {}^ae^{\nu\alpha} + \dots.$$
$$(16.1.12)$$

This complexity goes with a single vertex, which is always one half of an amplitude; when we compound these complexities, as for example in computing a diagram such as is shown in Figure 16.2(a), we may get as many as 108 terms.

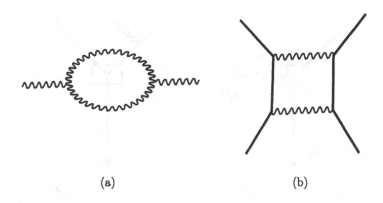

(a)   (b)

Figure 16.2

16.2   COMPLETION OF THE THEORY: A SIMPLE EXAMPLE OF GRAVITATIONAL RADIATION

In the preceding section we have given a complete outline of the theory. All that remains is to proceed to calculate the relevant diagrams in any physical process, with the same rules that are used in electrodynamics. Specific examples of some of the simplest diagrams were worked out in Lecture 4; for example, the amplitude for scattering by exchange of a single graviton is given in eq.(4.3.5). In practice, some care must be taken in properly symmetrizing some expressions, but this becomes easy with a little practice, and the abbreviations of the bar are useful in avoiding a lot of algebra.

In the lowest order, the theory is complete by this specification. All processes suitably represented by "tree" diagrams have no difficulties. The "tree" diagrams are those which contain neither bubbles nor closed loops of the kind shown in Figure 16.2. The name evidently refers to the fact that the branches of a tree never close back upon themselves.

In higher orders, when we allow bubbles and loops in the diagrams, the theory is unsatisfactory in that it gets silly results. Methods for curing this disease have been successful in curing only one-ring difficulties. In order to discuss the cures, it shall be easier for us to study briefly the Yang-Mills vector meson theory, which reproduces the same difficulties but is much easier to work with. Some of these difficulties have to do with the lack of unitarity of some sums of diagrams. We shall discuss a group of relations which hold between different kinds of diagrams. These have no direct tests in experiments on gravity, but some of them are familiar from work on other field theories.

I do not know whether it will be possible to develop a cure for treating the multi-ring diagrams. I suspect not—in other words, I suspect that the

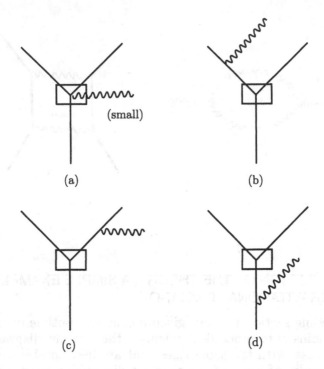

Figure 16.3

theory is not renormalizable. Whether it is a truly significant objection to a theory, to say that it is not renormalizable, I don't know.

The most interesting of the problems we shall take is perhaps that of the radiation of gravitational waves. As a beginning example, let us consider the radiation of a single graviton following the disintegration of some suitable particles. Since we shall be using a scalar theory of matter, perhaps it is best that we think of some decay of scalar particles, such as $K \to 2\pi$. The emission of a low-frequency graviton is necessary, in order to tell the outside world gravitationally that the decay has occurred, much in the way that a low energy photon must be emitted in a similar decay, since some charge has been accelerated. Of the many diagrams which may be written, that in which the graviton comes out of the vertex of the decay is usually much smaller, so we need not consider it at first. As an exercise, it might be useful to work out the last three diagrams shown in Figure 16.3.

16.3 RADIATION OF GRAVITONS WITH PARTICLE DECAYS

The coupling of gravitons to matter is so weak that there is truly no hope of observing quantum gravitational effects associated with particle events. In this sense, the calculations we are about to do are absolutely

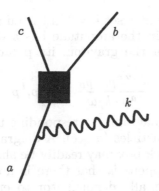

Figure 16.4

irrelevant in practice. Still, we have proposed a definite theory, and these unobservable processes are the simplest effects that our theory predicts; they would be observable and important if the coupling were stronger.

There are many one-graviton diagrams in the decay of a particle. For the illustration we take that in Figure 16.4, $a \to b + c$. The amplitude at the a-graviton vertex is given by

$$-2\lambda e^{\mu\nu}\left[{}^1p_\mu{}^2p_\nu - \frac{1}{2}\eta_{\mu\nu}\left({}^1p_\alpha{}^2p^\alpha - m^2\right)\right], \qquad (16.3.1)$$

where the preceding superscripts 1 and 2 denote the matter particle before and after the vertex. After the emission, the particle a propagates with momentum $({}^ap - k)$ to the decay vertex, hence ${}^2p^\alpha = ({}^ap - k)^\alpha$. If we let the decay amplitude be represented by a quantity A depending on the momenta of the three particles (a, b, c) whose trajectories enter the black box, the expression for the amplitude is

$$-2\lambda e^{\mu\nu}\left[{}^ap_\mu({}^ap - k)_\nu - \frac{1}{2}\eta_{\mu\nu}\left({}^ap \cdot ({}^ap - k) - m_a^2\right)\right]$$
$$\frac{1}{({}^ap - k)^2 - m_a^2}\left[A({}^ap - k, {}^bp, {}^cp)\right]. \qquad (16.3.2)$$

For our purposes, the exact nature of the amplitude A is unimportant; it represents whatever would have been there without the graviton.

The amplitude, eq.(16.3.2), is large only when the propagator has a very small value, that is, when k is very much smaller than ${}^ap$, so the propagation corresponds to a nearly free particle. In the limiting case of weak gravitons, the process is identical to that of braking radiation, "bremsstrahlung" emission of weak photons; it is also closely related to the classical limit, since it depends on how the charge (mass) currents move. The denominator is $-2\,{}^ap \cdot k$, and in the limit that the frequencies

ω of k are very small, we may set $k = 0$ in the numerator. If we factor out λ/ω, the second factor in the amplitude has a definite limit depending only on the direction of the graviton, its polarization, and the decay amplitude.

$$\frac{\lambda}{\omega} \cdot \frac{e^{\mu\nu}\,{}^a p_\nu\,{}^a p_\mu}{{}^a p \cdot k/\omega} \cdot A({}^a p, {}^b p, {}^c p) \tag{16.3.3}$$

There are three similar diagrams, corresponding to emission of a graviton from any of the three particles (a, b, c). A diagram corresponding to the graviton leaving the black box may readily be shown to be much smaller in magnitude; what happens is that there is no nearly free particle to propagate, hence no "small" denominator to enhance the term. If we neglect this term and all higher orders, we find that the amplitude to emit a number of gravitons is

$$\frac{\lambda}{\omega} \cdot a \cdot A({}^a p, {}^b p, {}^c p); \tag{16.3.4}$$

$$a = \sum_i {}^i p_\mu\,{}^i p_\nu (-)_i \frac{e^{\mu\nu}}{{}^i p \cdot (k/\omega)},$$

where i represents the particle joined to the graviton vertex and where $(-)_i$ is a factor which is $+1$ for an incoming particle and -1 for an outgoing particle. The quantity a is a kinematic and geometric factor. To compute a transition rate we square the amplitude, insert a density-of-state factor $k^2\,dk\,d\Omega/(2\pi)^3$, and a normalization factor which is $\pi/(2E_i)$ where E_i is the energy of each particle. The result is

$$P = a^2 \frac{d\Omega}{4\pi} \frac{d\omega}{\omega} \frac{\lambda^2}{4\pi^2}, \tag{16.3.5}$$

giving the probability of graviton emission per disintegration. The factor λ^2 makes the rate extremely low, so low that the odds are very much against there being a measurable graviton recoil in any cloud chamber or hydrogen chamber or spark chamber decay event ever recorded. The inverse energy factor $1/\omega$ makes the quantity large for extremely low graviton energies; however, the fact is almost irrelevant, since λ^2/ω becomes near 1 only for energies so low that the wavelength of the graviton would exceed the radius of the universe by some factor such as 10^{39}.

Although we have worked out the theory assuming scalar particles, in the low-energy limit the answer is the same no matter what the spin of the particles may be. This is because in the low energy limit only the mass currents, the movement of the masses is relevant. Our answer, of course, has an infrared divergence, so that the probability for the emission of a graviton, (if its energy is irrelevant) appears infinite. The trouble is no more serious than the infrared divergence for the emission of low energy photons—and the troubles may be cured by the same tricks as in low-energy "bremsstrahlung."

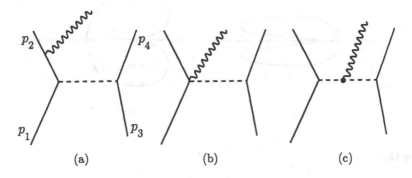

Figure 16.5

16.4   RADIATION OF GRAVITONS WITH PARTICLE SCATTERING

A soft graviton may be emitted when two particles scatter by any process, including graviton exchange. The first order diagrams which are involved are of the three types shown in Figure 16.5. In the low-energy limit, only the diagrams of type (a), that is, in which the graviton vertex joins a free particle, are important. The other two are processes much less probable if the graviton momentum k is much smaller than the momentum transfer q; in (b), for example, there is no almost-free particle which propagates, hence no small denominator. In (c), the second propagator is of order $1/(q-k)^2 \approx 1/q^2$. As far as the radiation is concerned, the exact nature of the over-all scattering process is not important. I emphasize the last point because there are always some theorists who go about mumbling some mystical reasons to claim that the radiation would not occur if the scattering is gravitational—there is no basis for these claims; as far as we are concerned, radiation of gravity waves is as real as can be; the sun-earth rotation must be a source of gravitational waves. Actually, we should perhaps in this section restrict our thoughts to particle scattering; for the motion of big objects such as planets or stars it may be more consistent to work in the classical limit. Gravity is not always negligible—only in processes of atomic collisions.

The structure of the four amplitudes corresponding to diagrams such as Figure 16.5(a) is the same as in the particle decays. If we describe the graviton polarization by a tensor e, the total amplitude is proportional to the scattering amplitude for no graviton, to some energy factors, and to the quantity

$$a = \sum_i (-)_i \frac{p \cdot e \cdot {}^i v}{1 - {}^i v \cos \theta}. \qquad (16.4.1)$$

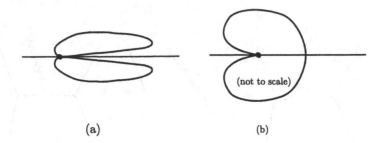

(a)   (b)

Figure 16.6

The denominator represents the product $^ip \cdot k = E\omega - \vec{p} \cdot \vec{k}$ when two energies E and ω have been factored out. Since $\vec{p} = E\vec{v}$, the superscripts i refer to each of the four particle branches in the collision. The numerator contains the contracted product of the polarization tensor $e_{\mu\nu}$ with the two momenta (before and after) of the particle it couples—this is the only physical second-rank tensor which may be constructed for a scalar particle.

The answers we have obtained are very closely analogous to those occurring in photon emission: the essential difference consists in that electromagnetism couples to a vector whereas gravity couples to a tensor. For a fast-moving particle, $v \approx c$, so that the denominator in eq.(16.4.1) can be very small and the amplitude could become very large near $\theta_i \approx 0$. On the other hand, the polarization tensor is always transverse to the graviton momentum. In electromagnetism, a vector polarization is also transverse to the photon momentum; there is only one dot product in the numerator so that, when θ is small and $v \approx c$,

$$a_{\text{e.m.}} \rightarrow \frac{p \cdot e}{1 - v \cos\theta} \propto \frac{\sin\theta}{1 - \cos\theta} \approx \frac{2}{\theta}. \tag{16.4.2}$$

The photon emission can become very large for small angles. It does not actually blow up because v is never quite equal to c. The radiation intensity pattern corresponding to a single particle has two lobes as shown in Figure 16.6. In gravity, the coupling is tensor, so that it is doubly transverse; in the limit $\theta \rightarrow 0$ and $v = 1$,

$$a_{\text{g}} \rightarrow \frac{p \cdot e \cdot p}{1 - v \cos\theta} \propto \frac{\sin^2\theta}{1 - v \cos\theta} \approx 2, \tag{16.4.3}$$

so that the intensity pattern is not strongly forward, but on the whole rather uniform in comparison (Figure 16.6(b)). The difference may perhaps be viewed intuitively as being due to the fact that, in being produced, a spin of two requires more "transversality" than a spin of one.

There is one amplitude having an angular structure such as in Figure 16.6(b) about each of the four directions of the particles in the scattering problem. The graviton emission intensity is the square of the sum of the four amplitudes, so that in general it will be rather symmetric.

For slow-moving particles, $v \ll c$, the denominator plays essentially no role and the pattern is determined solely by the numerator. This may be expressed as the contracted product of two tensors.

$$\sum_i (-)_i \, {}^i p \cdot e \cdot {}^i v = e^{\alpha\beta} S_{\alpha\beta}, \qquad \alpha, \beta = x, y, z,$$

$$S_{\alpha\beta} = \sum_i (-)_i \, p_\alpha v_\beta \tag{16.4.4}$$

The character of the radiation is determined entirely by the tensor $S_{\alpha\beta}$ which represents the stress produced in the collision. We recognize that its form is precisely analogous to that of a stress in a moving fluid.

$$\text{Stress} = \rho v_\alpha v_\beta \qquad (\rho = \text{mass density}) \tag{16.4.5}$$

If we have a collision between two particles, the stress $S_{\alpha\beta}$ has a simple expression in terms of the average velocities (before and after) of the colliding particles. We let the momentum transfer be $Q = p_2 - p_1 = -p_4 + p_3$. (See Figure 16.5.) Write down the average velocities.

$$v = (p_2 + p_1)/2m, \qquad v' = (p_3 + p_4)/2m' \tag{16.4.6}$$

In terms of these combinations, it may readily be shown after proper symmetrization that

$$S_{\alpha\beta} = 2(v' - v)_\alpha Q_\beta. \tag{16.4.7}$$

With this formula we can now answer an interesting question; in a collision between a light and a heavy particle, which one does most of the radiating? The formula tells us that if $v' \ll v$, the radiation depends only on v. In considering radiation from glancing collisions of a very light particle with a massive object, we now know for sure that it is legitimate to consider the massive particle to be always at rest. This rule works provided the acceleration is nearly perpendicular to the velocity, so $\vec{Q} \cdot \vec{v} \approx 0$.

The formulae here apply for both elastic and inelastic collisions, which can leave one or both of the masses in excited states.

16.5 THE SOURCES OF CLASSICAL GRAVITATIONAL WAVES

We now pass on to a description of classical gravitational radiation. Just as in the quantum-mechanical case, we shall find that the emitter of radiation is also the stress. The starting point is the differential equation

$$\Box^2 \overline{h}_{\mu\nu} = \lambda S_{\mu\nu}. \tag{16.5.1}$$

The solution proceeds exactly as in electrodynamics for the solution of the vector potentials produced by arbitrary currents. If we assume a harmonic time variation as $\exp(-i\omega t)$ for all quantities, the vector potential is given by

$$A_\mu(1) = \int dV_2 \frac{j_\mu(2) \cdot \exp(i\omega r_{12})}{4\pi r_{12}}, \tag{16.5.2}$$

where the indices 1 and 2 refer to different space positions; (1) is the place at which we compute the potentials A_μ, (2) are the places at which the currents are and r_{12} is their separation. One of the simplest cases of radiation corresponds to an oscillating dipole, such that the currents are confined to a small region of space. It is fairly straightforward to compute the space components $A_x \, A_y \, A_z$; the time component, or scalar potential, is ordinarily most easily obtained from the divergence condition on A_μ.

$$A^\mu{}_{,\mu} = 0 \longrightarrow i\omega A_t = \nabla \cdot \vec{A} \tag{16.5.3}$$

The situation is precisely analogous in gravitation. The time parts of the fields $\overline{h}_{\mu\nu}$ are most easily obtained from the divergence conditions after computing the space parts by the following rule.

$$\overline{h}_{\mu\nu}(1) = -\frac{\lambda}{4\pi} \int dV_2 \, S_{\mu\nu}(2) \frac{\exp(i\omega r_{12})}{r_{12}}. \tag{16.5.4}$$

To compute such things as the power radiated away, we consider that the point (1) is far away, at some distance much greater than the dimensions of the region where $S_{\mu\nu}(2)$ is expected to be large as illustrated in Figure 16.7. We may expand the distance r_{12} as a power series in the radial distances of (1) and (2) from some origin near the points (2), and we find

$$r_{12} = \sqrt{r_1^2 + r_2^2 - 2r_1 r_2 \cos\theta} = r_1\sqrt{1 - (2r_2/r_1)\cos\theta + \cdots} \tag{16.5.5}$$
$$\approx r_1 - r_2 \cos\theta + \cdots,$$

when $r_2 \ll r_1$. Here $\cos\theta$ is the angle between r_2 and r_1. Since any waves observed at (1) will have a momentum vector directed along r_1, we obtain the following expression for $\overline{h}_{\mu\nu}(1)$.

$$\overline{h}_{\mu\nu}(1) = -\frac{\lambda}{4\pi r_1} e^{i\omega r_1} \int d^3 r_2 \, S_{\mu\nu}(2) \, e^{-i\vec{K}\cdot\vec{r}_2} \tag{16.5.6}$$

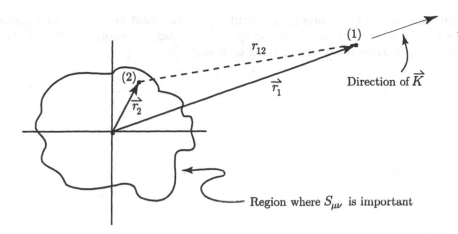

Figure 16.7

The integral appearing in eq.(16.5.6) is now independent of the point (1); we see that the stress tensor $S_{\mu\nu}(2)$ is a source of spherical waves.

In the case of electromagnetism, the simplest cases of radiation often correspond to the dipole approximation, which is the first nonvanishing term in the sequence of integrals corresponding to an expansion of the exponential. Because the source is a tensor rather than a vector, the first nonvanishing term in gravitation is of a quadrupole character. The expansion is justified if the frequencies are such that $\vec{K} \cdot \vec{r}_2$ is smaller than 1 over the region where $S_{\mu\nu}$ is significant. For all rotating masses such as double stars or star-planet systems, the periods of the motion (~ 1 year for earth-sun system, say) are much longer than the time that gravity takes to propagate through a distance of the order of the size of the system (~ 16 minutes for the earth-sun system), so that the terms of the expansion get smaller very rapidly. Thus, in nearly all cases of astronomical interest, wavelengths are much longer than the object dimensions. The result is that the fields $\bar{h}_{\mu\nu}$ are proportional to the integrals of the transverse stresses (the total transverse stress).

$$\bar{h}_{ab} = -\lambda \frac{e^{i\omega r}}{4\pi R} S_{ab} \quad \text{where} \quad S_{ab} = \int d^3 r\, S_{ab}(\vec{r}) \qquad (16.5.7)$$

The stresses in a direction *along* the wave vector are irrelevant. Every qualitative rule that was useful in electromagnetism is carried bodily over into gravitation.

What is the power radiated by such a wave? There are a great many people who worry needlessly at this question, because of a perennial prejudice that gravitation is somehow mysterious and different—they feel that it might be that gravity waves carry no energy at all. We can definitely show that they can indeed heat up a wall, so there is no question as to their energy content. The situation is exactly analogous to electrodynamics—and in the quantum interpretation, every radiated graviton carries away an amount of energy $\hbar\omega$.

Bibliography

[Alva 89] Alvarez, Enrique (1989). "Quantum gravity: an introduction to some recent results," *Reviews of Modern Physics*, **61**, 561–604.

[Asht 86] Ashtekar, A. (1986). "New variables for classical and quantum gravity," *Physical Review Letters*, **57**, 2244–2247.

[Asht 87] Ashtekar, A. (1987). "A new Hamiltonian formulation of general relativity," *Physical Review D*, **36**, 1587–1603.

[Baad 52] Baade, Walter (1952). "Report of the Commission on Extragalactic Nebulae," *Transactions of the International Astronomical Union*, **8**, 397–399.

[Bard 65] Bardeen, James M. (1965). "Stability and dynamics of spherically symmetric masses in general relativity," unpublished Ph.D. thesis, California Institute of Technology.

[BBIP 91] Balbinot, Roberto, Brady, Patric R., Israel, Werner, and Poisson, Eric (1991). "How singular are black-hole interiors?" *Physics Letters A*, **161**, 223–226.

[Birk 43] Birkhoff, G. (1943). "Matter, electricity, and gravitation in flat space-time," *Proc. Nat. Acad. Sci. U.S.*, **29**, 231–239.

[BTM 66] Bardeen, James M., Thorne, Kip S., and Meltzer, David W. (1966). "A catalog of methods for studying the normal modes of radial pulsation of general relativistic stellar models," *Astrophysical Journal*, **145**, 505–513.

[Bond 57] Bondi, Hermann, "Plane gravitational waves in general relativity," *Nature*, **179**, 1072–1073.

[BoDe 75] Boulware, David G. and Deser, Stanley (1975). "Classical general relativity derived from quantum gravity," *Annals of Physics*, **89**, 193–240.

[Cart 28] Cartan, Elie (1928). *Leçons sur la Géométrie des Espaces de Riemann*, Memorial des Sciences Mathématiques, Fascicule IX (Gauthier-Villars, Paris, France).

[Chan 64] Chandrasekhar, S. (1964). "Dynamical instability of gaseous masses approaching the Schwarzschild limit in general relativity," *Physical Review Letters*, **12**, 114–116.

[ChHa 82] Chandrasekhar, S. and Hartle, James B. (1982). "On crossing the Cauchy horizon of a Reissner-Nordström black hole," *Proceedings of the Royal Society of London A*, **384**, 301–315.

[Cock 65] Cocke, W. John (1965). "A maximum entropy principle in general relativity and the stability of fluid spheres," *Annales de l'Institut Henri Poincaré*, **A 2**, 283–306.

[Dese 70] Deser, Stanley (1970). "Self-interaction and gauge invariance," *General Relativity and Gravitation*, **1**, 9–18.

[Dese 87] Deser, Stanley (1987). "Gravity from self-interaction in a curved background," *Classical and Quantum Gravity*, 4, L99-L105.

[DeWi 57] DeWitt, Cecile M. (1957). *Conference on the Role of Gravitation in Physics* at the University of North Carolina, Chapel Hill, March 1957; WADC Technical Report 57-216 (Wright Air Development Center, Air Research and Development Command, United States Air Force, Wright Patterson Air Force Base, Ohio).

[DeWi 67a] DeWitt, Bryce S. (1967). "Quantum theory of gravity, II," *Physical Review*, **162**, 1195–1239.

[DeWi 67b] DeWitt, Bryce S. (1967). "Quantum theory of gravity, III," *Physical Review*, **162**, 1239–1256.

[DeWi 94] DeWitt, Bryce S. (1994). Private communication.

[Dira 37] Dirac, P. A. M. (1937). "The cosmological constants," *Nature*, **139**, 323.

[Dira 38] Dirac, P. A. M. (1938). "New basis for cosmology," *Proc. R. Soc. London A*, **165**, 199–208.

[DrMa 77] Drechsler, W. and Mayer, M. E. (1977). *Fiber Bundle Techniques in Gauge Theories* (Lecture Notes in Physics, Volume 67, Springer-Verlag, New York).

[Eddi 31] Eddington, A. (1931). "Preliminary note on the masses of the electron, the proton, and the universe," *Proc. Cambridge Phil. Soc.*, **27**, 15–19.

[Eddi 36] Eddington, A. (1936). *Relativity Theory of Protons and Electrons* (Cambridge University Press, Cambridge).

[Eddi 46] Eddington, A. (1946). *Fundamental Theory* (Cambridge University Press, Cambridge).

[Eins 39] Einstein, Albert (1939). "On a stationary system with spherical symmetry consisting of many gravitating masses," *Annals of Mathematics*, **40**, 922-936.

[FaPo 67] Faddeev, L. D. and Popov, V. N. (1967). "Feynman diagrams for the Yang-Mills Field," *Physics Letters B*, **25**, 29–30.

[Feyn 57] Feynman, Richard P. (1957). "Conference on the Role of Gravitation in Physics, an expanded version of the remarks by R. P. Feynman on the reality of gravitational waves, mentioned briefly on page 143 of the Report [DeWi 57]," typescript in Box 91, File 2 of The Papers of Richard P. Feynman, the Archives, California Institute of Technology.

[Feyn 61] Feynman, Richard P. (1961). Unpublished letter to Victor F. Weisskopf, January 4–February 11, 1961; in Box 3, File 8 of The Papers of Richard P. Feynman, the Archives, California Institute of Technology.

[Feyn 63a] Feynman, Richard P., Leighton, Robert B., and Sands, Matthew (1963). *The Feynman Lectures on Physics* (Addison-Wesley, Reading, Massachusetts).

[Feyn 63b] Feynman, Richard P. (1963). "Quantum theory of gravitation," *Acta Physica Polonica*, **24**, 697–722.

[Feyn 67] Feynman, Richard P. (1967). *The Character of Physical Law* (M.I.T., Cambridge).

[Feyn 72] Feynman, Richard P. (1972). "Closed loop and tree diagrams" and "Problems in quantizing the gravitational field, and the massless Yang-Mills field," in *Magic Without Magic: John Archibald Wheeler*, edited by John R. Klauder (W. H. Freeman, San Francisco), pp. 355–408.

[Feyn 85] Feynman, Richard P., as told to Leighton, Ralph (1985). *Surely You're Joking, Mr. Feynman!* (W. W. Norton, New York).

[Feyn 88] Feynman, Richard P., as told to Leighton, Ralph (1988). *What Do You Care What Other People Think?* (W. W. Norton, New York).

[Feyn 89] "Feynman's office: The last blackboards," *Physics Today*, **42** (2), 88 (1989).

[FiPa 39] Fierz, M. and Pauli, W. (1939). "Relativistic wave equations for particles of arbitrary spin in an electromagnetic field," *Proceedings of the Royal Society of London A*, **173**, 211–232.

[Fink 58] Finkelstein, David (1958). "Past-future asymmetry of the gravitational field of a point particle," *Physical Review*, **110**, 965–967.

[Fowl 64] Fowler, William A. (1964). "Massive stars, relativistic polytropes, and gravitational radiation," *Reviews of Modern Physics*, **36**, 545–555.

[FuWh 62] Fuller, Robert W. and Wheeler, John A. (1962). "Causality and multiply connected space-time," *Physical Review*, **128**, 919–929.

[FWML 74] Fairbank, W., Witteborn, F., Madey, J., and Lockhart, J. (1974). "Experiments to determine the force of gravity on single electrons and positrons," *Experimental Gravitation: Proceedings of the International School of Physics "Enrico Fermi,"* Course LVI, B. Bertotti, ed. (Academic Press, New York) 310–330.

[Gell 89] Gell-Mann, Murray (1989). "Dick Feynman—The guy in the office down the hall," *Physics Today*, **42** (2), 50-54.

[Good 61] Good, M. L. (1961). "K_2^0 and the equivalence principle," *Physical Review,* **121**, 311–313.

[GoSa 86] Goroff, M. and Sagnotti, A. (1986). *Nuclear Physics B*, **266**, 709.

[GrBr 60] Graves, John C. and Brill, Dieter R. (1960). "Oscillatory character of Reissner–Nordström metric for an ideal charged wormhole," *Physical Review*, **120**, 1507–1513.

[GSW 87] Green, M., Schwarz, J., and Witten, E. (1987). *Superstring theory* (Cambridge University Press, Cambridge, England).

[GrPe 88] Gross, D. and Periwal, V. (1988). "String perturbation theory diverges," *Physical Review Letters*, **60**, 2105–2108.

[Gupt 54] Gupta, Suraj N. (1954). "Gravitation and electromagnetism," *Physical Review*, **96**, 1683–1685.

[Guth 81] Guth, Alan H. (1981). "Inflationary universe: a possible solution to the horizon and flatness problems," *Physical Review D*, **23**, 347–356.

[HaEl 73] Hawking, Stephen W. and Ellis, George F. W. (1973). *The Large Scale Structure of Space-time* (Cambridge University Press, Cambridge, England).

[Hatf 92] Hatfield, B. (1992). *Quantum Field Theory of Point Particles and Strings* (Addison-Wesley, Reading, MA).

[Hilb 15] Hilbert, D. (1915). "Die Grundlagen der Physik," *Konigl. Gesell. d. Wiss. Göttingen, Nachhr. Math.-Phys. Kl.*, 395–407.

[HoFo 63] Hoyle, Fred, and Fowler, William A. (1963). "On the nature of strong radio sources," *Monthly Notices of the Royal Astronomical Society*, **125**, 169–176.

[Hoyl 48] Hoyle, Fred (1948). "A new model for the expanding universe," *Mon. Not. R. Astron. Soc.*, **108**, 372–382.

[HWWh 58] Harrison, B. Kent, Wakano, Masami, and Wheeler, John A. (1958). "Matter-energy at high density; endpoint of thermonuclear evolution," in Onzième Conseil de Physique Solvay, *La Structure et l'Evolution de l'Univers* (Editions R. Stoops, Brussels), pp. 124–148.

[Iben 63] Iben, Icko, Jr. (1963). "Massive stars in quasistatic equilibrium," *Astrophysical Journal*, **138**, 1090–1096.

[JaSm 88] Jacobson, T. and Smolin, L. (1988). "Nonperturbative quantum geometries," *Nuclear Physics B*, **299**, 295–345.

[JeDG 53] Jennison, R. C. and Das Gupta, M. K. (1953). "Fine structure of the extra-terrestrial radio source Cygnus 1," *Nature*, **172**, 996–997.

[Klei 89] Kleinert, H. (1989). "Quantum mechanics and path integrals in spaces with curvature and torsion," *Modern Physics Letters A*, **4**, 2329–2337.

[Kore 74] Korepin, Vladimir (1974). Unpublished Diplom thesis, Leningrad State University.

[Krai 47] Kraichnan, Robert H. (1947). "Quantum theory of the linear gravitational field," unpublished B.S. thesis, Massachusetts Institute of Technology.

[Krai 55] Kraichnan, Robert H. (1955). "Special relativistic derivation of generally covariant gravitation theory," *Physical Review*, **98**, 1118–1122.

[Krai 56] Kraichnan, Robert H. (1956). "Possibility of unequal gravitational and inertial masses," *Physical Review*, **101**, 482–488.

[Krus 60] Kruskal, Martin (1960). "Maximal extension of the Schwarzschild metric," *Physical Review*, **119**, 1743–1745.

[LaLi 51]   Landau, L. D. and Lifshitz, E. M. (1951). *The Classical Theory of Fields*, translated by M. Hammermesh (Addison-Wesley, Reading, Massachusetts).

[LiBr 90]   Lightman, Alan and Brawer, Roberta (1990). *Origins: The Lives and Worlds of Modern Cosmologists* (Harvard University Press, Cambridge).

[MaWh 66]   May, Michael M. and White, Richard H. (1966). "Hydrodynamic calculations of general relativistic collapse," *Physical Review*, **141**, 1232–1241.

[Mich 63]   Michael, F. Curtis (1963). "Collapse of massive stars," *Astrophysical Journal*, **138**, 1097–1103.

[Miln 34]   Milne, E. (1934). *Quart. J. Math (Oxford)*, **5**, 64; McCrea, W. and Milne, E. (1934). *Quart. J. Math (Oxford)*, **5**, 73.

[MiWh 57]   Misner, Charles W. and Wheeler, John A. (1957). "Classical physics as geometry: gravitation, electromagnetism, unquantized charge, and mass as properties of curved empty space," *Annals of Physics*, **2**, 525–603, reprinted in [Whee 62].

[MTW 73]   Misner, Charles W., Thorne, Kip S., and Wheeler, John A. (1973). *Gravitation* (W. H. Freeman, San Francisco).

[NiGo 91]   Nieto, M. and Goldman, T. (1991). "The arguments against "antigravity" and the gravitational acceleration of antimatter," *Physics Reports*, **205**, 221–281.

[OpSn 39]   Oppenheimer, J. Robert and Snyder, Hartland (1939). *Physical Review*, **56**, 455–459.

[Penr 65]   Penrose, Roger (1965). "Gravitational collapse and space-time singularities," *Physical Review Letters*, **14**, 57–59.

[Podu 64]   Podurets, Mikhail A. (1964). "The collapse of a star with back pressure taken into account," *Doklady Akademi Nauk*, **154**, 300-301; English translation, *Soviet Physics—Doklady*, **9**, 1–2.

[RoSm 88]   Rovelli, C. and Smolin, L. (1988). "Loop representation for quantum general relativity," *Nuclear Physics B*, **331**, 80–152.

[Schi 58]   Schiff, L. (1958). "Sign of the gravitational mass of the positron," *Physical Review Letters*, **1**, 254–255.

[Schi 59]   Schiff, L. (1959). "Gravitational properties of antimatter," *Proc. Natl. Acad. Sci.*, **45**, 69–80.

[Schm 63]   Schmidt, Maarten A. (1963). "3C 273: A star-like object with large red-shift," *Nature*, **197**, 1040.

[Schu 85] Schutz, Bernard F. (1985). *A First Course in General Relativity* (Cambridge University Press, Cambridge).

[Schw 16] Schwarzschild, Karl (1916). "Über das Gravitationsfeld eines Massenpunktes nach der Einsteinschen Theorie," *Sitzungsberichte der Deutschen Akademie der Wissenschaften zu Berlin, Klasse für Mathematik, Physik, und Technik*, **1916**, 189–196.

[Syke 94] Sykes, Christopher (1994). *No ordinary genius: the illustrated Richard Feynman* (W. W. Norton, New York).

[Thir 18] Thirring, H. (1918). "Über die Wirkung rotierender ferner Massen in der Einsteinschen Gravitationstheorie," *Phys. Z.*, **19**, 33–39; Thirring, H. and Lense, J. (1918). "Über den Einflußder Eigenrotation der Zentralkörper auf die Bewegung der Planeten und Monde nach der Einsteinschen Gravitationstheorie," *Phys. Z.*, **19**, 156–163; Thirring, H. (1921). "Berichtigung zu meiner Arbeit: 'Über die Wirkung rotierender ferner Massen in der Einsteinschen Gravitationstheorie'," *Phys. A.*, **22**, 29–30.

[tHVe 74] 't Hooft, G. and Veltman, M. (1974). *Ann. Inst. Henri Poincaré*, **20**, 69.

[Thor 94] Thorne, Kip S. (1994). *Black Holes and Time Warps: Einstein's Outrageous Legacy* (W. W. Norton, New York).

[Toop 66] Tooper, Robert F. (1966). "The 'standard model' for massive stars in general relativity," *Astrophysical Journal*, **143**, 465–482.

[vanN 81] van Nieuwenhuizen, Peter (1981). "Supergravity," *Physics Reports*, **68**, 189–398.

[Vebl 27] Veblen, O. (1927). *Invariants of Quadratic Differential Forms* (Cambridge Tracts in Math and Math Phys., Cambridge University Press, London).

[Wald 84] Wald, Robert M. (1984). *General Relativity* (University of Chicago Press, Chicago).

[Wald 86] Wald, Robert M. (1986). "Spin-two fields and general covariance," *Physical Review*, **D33**, 3613–3625.

[Wein 64a] Weinberg, Steven (1964). "Derivation of gauge invariance and the equivalence principle from Lorentz invariance of the S-matrix," *Physics Letters*, **9**, 357–359.

[Wein 64b] Weinberg, Steven (1964). "Photons and gravitons in S-matrix theory: derivation of charge conservation and equality of gravitational and inertial mass," *Physical Review,* **135**, B1049 –B1056.

[Wein 72] Weinberg, Steven (1972). *Gravitation and Cosmology* (Wiley, New York).

[Wein 79] Weinberg, Steven (1979). "Ultraviolet divergences in quantum theories of gravitation," in *General Relativity: An Einstein Centenary Volume*, edited by S. W. Hawking and W. Israel (Cambridge University Press, Cambridge, England), pp. 790–831.

[Wein 89] Weinberg, Steven (1989). "The cosmological constant problem," *Reviews of Modern Physics*, **61**, 1-22.

[Went 49] Wentzel, G. (1949). *Quantum Theory of Fields* (Interscience, New York).

[Whee 62] Wheeler, John A. (1962). *Geometrodynamics* (Academic Press, New York).

[Whee 90] Wheeler, John A. (1990). *A Journey into Gravity and Spacetime* (Scientific American Library, W. H. Freeman, New York).

[WhFe 45] Wheeler, John A. and Feynman, Richard P. (1945). "Interaction with the absorber as the mechanism of radiation," *Reviews of Modern Physics*, **17**, 157–181.

[WhFe 49] Wheeler, John A. and Feynman, Richard P. (1949). "Classical electrodynamics in terms of direct interparticle action," *Reviews of Modern Physics*, **21**, 425–433.

[WiFa 67] Witteborn, F. and Fairbank, W. (1967). "Experimental comparison of the gravitational force on freely falling electrons and metallic electrons," *Physical Review Letters*, **19**, 1049–1052.

[Yang 77] Yang, C. N. (1977). *Ann. New York Acad. Sci.*, **294**, 86.

Index

Printed in the United States
by Baker & Taylor Publisher Services

Printed in the United States
by Baker & Taylor Publisher Services